Quark-Gluon Plasma: A Recreated Primordial Soup

Sartre

Contents

4 Tsallis fits with radial flow

5 Tsallis fits with chemical potential

6 Tsallis fits to Pb-Pb collisions

III Applications of extensive statistics 75

7 Basic literature

8 Thermal model fits

IV   Discussion and Conclusion   **87**

Chapter 1

Introduction to high energy collisions

Atoms consist of protons and neutrons: inside these, there exist quarks and their interaction are mediated by gluons. One way to study the fundamental building blocks of nuclear matter is to set free these fundamental particles. Through high energy nuclear collisions, we attain high enough centre of mass energies to liberate these fundamental particles inside a quark-gluon plasma (QGP).

These nuclear collisions are conducted at various laboratories around the world e.g. the Relativistic Heavy Ion Collider (RHIC) at the Brookhaven National Laboratory, the Large Hadron Collider (LHC) in Geneva, the Nuclotron-based Ion Collider Facility (NICA) in Dubna and the Facility for Antiproton and Ion Research (FAIR) in Germany, just to mention a few.

The LHC is currently the largest particle accelerator in the world. It was build to provide some insights into some of the fundamental questions in particle physics. The experiments of interest here are: A Large Ion Collider Experiment (ALICE), the Compact Muon Solenoid (CMS) experiment and A Toroidal LHC Apparatus (ATLAS) experiment as well as the NA49 experiment and the RHIC accelerator.

The schematic layout of the Large Hadron Collider at CERN is presented in Figure 1.1, where beams of either protons p and lead Pb particles from the linear accelerators are first injected into the proton synchrotron (PS) then the super proton synchronous (SPS) accelerators and final into the 27 km LHC ring.

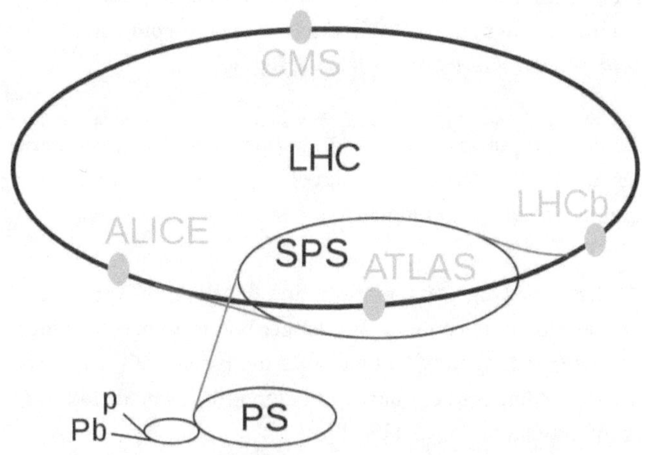

FIGURE 1.1: The schematic layout of the Large Hadron Collider
at CERN [6].

The ALICE, CMS, ATLAS and LHCb are the different detectors on the
ring. Most of the data we make use of has been made available by researchers
utilising the ALICE and CMS detectors.

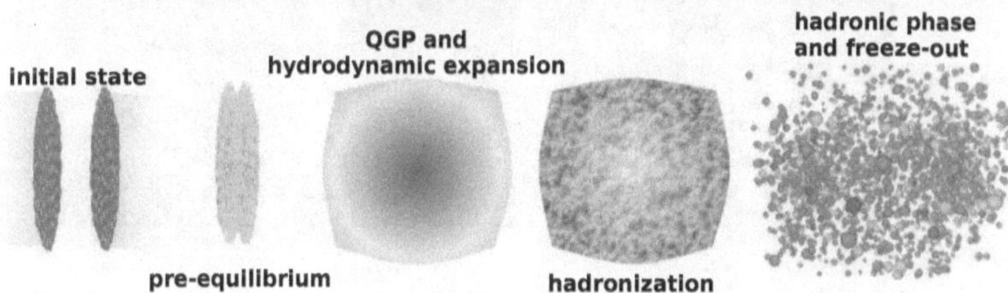

FIGURE 1.2: A depiction of a heavy ion collision [7]. Here we
show the system as it evolves in time from left to right, starting
from the initial state all the way to freeze-out.

Initial state: nuclei are accelerated at relativistic speeds (in excess of
99.99% of the speed of light) and they appear disk-like due to Lorentz
contraction.

- Pre - equilibrium: if the energy densities created during the collision are high enough, the collision produces a dense pre-equilibrium matter consisting of quarks, anti quarks and gluons.

As time progresses, the pre-equilibrium matter reaches thermal equilibrium leading to the formation of a system of deconfined quarks and gluons: the quark gluon plasma.

At these very high temperatures and densities, the quarks and gluons are said to be de-confined "no longer bound to hadron states", this is because the long range interactions are dynamically screened creating a system composed of quarks and gluons [8], as predicted by Quantum chromodynamics (QCD) [9, 10].

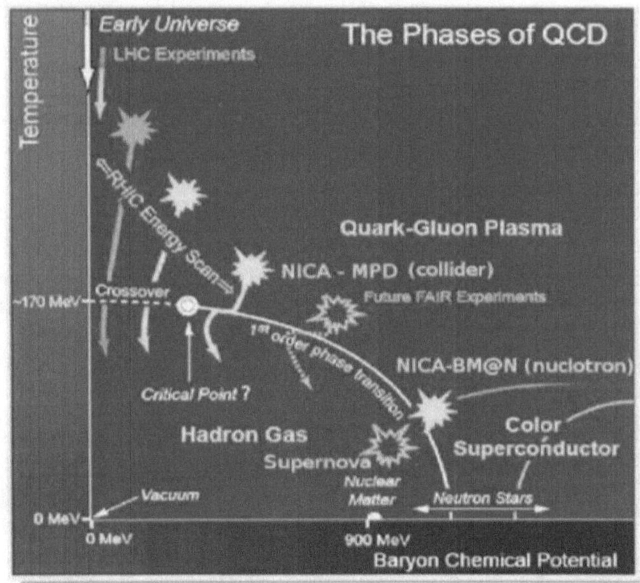

FIGURE 1.3: The quark gluon plasma (QGP) phase diagram; it presents the phases of the (QGP) as a function of temperature (T) and baryon chemical potential (μ_B) [11].

Figure 1.3 shows the phase diagram. At low temperature T and baryon chemical potential μ_B, the system is dominated by hadrons while at higher T or μ_B, the system is composed of mainly the quark gluon plasma, in simplest terms, QCD predicts that there has to be a phase transition from hadrons to QGP [12]. Higher temperatures are obtained through colliding nuclei at centre of mass energies greater than the rest mass of the nuclei.

- QGP and hydrodynamic expansion: as the system expands, after thermalization, the QGP is now driven by thermal gradients with the scattering lengths smaller than the system size. This expansion results in a decrease of the energy densities which become low to sustain deconfinement of partons and hadronization occurs [13, 14].

- Hadronic phase and freeze-out: as the system continues to expand, it becomes dilute and the individual hadrons free-stream to the detectors. An analysis of these hadrons reveals information about the QGP [13, 15].

1.2   High energy physics units

The Standard International (SI) measurement system are not appropriate for use in high energy physics calculations, this is because we deal mostly on microscopic scales of length, mass and time.

TABLE 1.1:   Some of the high energy units and their SI unit equivalent.

Quantity	High energy physics unit	SI Unit
Length	fm	$1\,\text{fm} = 10^{-15}\,\text{m}$
Mass	eV	$1\,\text{GeV}/c^2 = 1.78 \times 10^{-27}\,\text{kg}$
Energy	eV	$1\,\text{GeV} = 1.602 \times 10^{-10}\,\text{J}$
$\hbar$	1	$6.588 \times 10^{-25}\,\text{GeV s} = 1.055 \times 10^{-34}\,\text{J s}$
c	1	$2.998 \times 10^{23}\,\text{fm s}^{-1} = 2.998 \times 10^8\,\text{m s}^{-1}$
$\hbar c$	1	$197.5\,\text{MeV fm}$

Presented in Table 1.1 are the high energy units and their SI unit equivalent. We set the constants $\hbar = c = 1$ in our calculations: a consequence of this convenience is that the unit of mass and energy becomes the same.

1.3   Properties of quarks

The particles produced in heavy ion collisions are categorized in two classes: baryons with half-integer spin and obey Fermi-Dirac statistics; and mesons with a zero or integer spin and obey Bose-Einstein statistics. In Table 1.2, we present the composition of different quark flavours.

<TABLE 1.2: Properties of quarks: all quark flavours have spin 1/2 and baryon number +1/3. Here, Q is Electric Charge, S is Strangeness, C is Charm, b is beauty and T is Top.>

TABLE 1.2: Properties of quarks: all quark flavours have spin 1/2 and baryon number +1/3. Here, Q is Electric Charge, S is Strangeness, C is Charm, b is beauty and T is Top.

Flavour	Name	Mass (GeV)	Q	S	C	b	T
u	up	$0.0015 - 0.004$	$+2/3$	0	0	0	0
d	down	$0.004 - 0.008$	$-1/3$	0	0	0	0
t	top	174.3 ± 5.1	$+2/3$	0	0	0	$+1$
b	bottom	$4.1 - 4.4$	$-1/3$	0	0	$+1$	0
c	charm	$1.15 - 1.35$	$+2/3$	0	$+1$	0	0
s	strange	$0.080 - 0.130$	$-1/3$	-1	0	0	0

An example of a meson is a pion, it is made up of a combination of quark and antiquark $u\bar{d}$ pair. An example of baryon is the Lambda particle which is made up of an up quark, a down quark, and a strange quark. In Table 1.3, we present the composition of a few particle species that are mostly used in this book.

TABLE 1.3: A list of stable particles species considered in this book.

Type	Name	Mass (MeV)	Quark composition
π^-	antipion	139.57018 ± 0.00035	$\bar{u}d$
π^+	pion	139.57018 ± 0.0005	$u\bar{d}$
K^+	kaon	493.677 ± 0.013	$u\bar{s}$
K^-	antikaon	493.677 ± 0.013	$\bar{u}s$
p	proton	938.272081 ± 0.000006	uud
$\bar{p}$	antiproton	938.272081 ± 0.000006	$\bar{u}\bar{u}\bar{d}$
Λ^0	Lambda	1115.683 ± 0.006	uds
Ω^-	Omega	1672.45 ± 0.29	sss
Ξ^-	Xsi	1321.71 ± 0.07	dss

1.4   The role of resonances

Some of the particles produced in heavy ion collisions are resulting from the decay of resonances. When one makes use of the Thermal Model [16] to approximate particle yields, the daughter particles of resonances are added to the number of primary particles produced directly as a result of the collision.

1.5   Study objectives

High-energy collisions result in a large number of hadrons in the final state. To analyse the properties of such a large system of particles, we make use of statistical concepts. Several models are used to describe hadronic observables: these include the elliptic flow, transverse momentum spectra, rapidity distributions etc. The study objectives are in two parts given below:

1. The application of non-extensive statistics (Tsallis distribution) to describe the latest experimental results obtained at CERN using the Large Hadron Collider and other laboratories. The aim is to determine systematic behaviour of the parameters obtained and to gain theoretical insight in these.

   We also determine thermodynamic parameters: energy density, pressure, entropy density, temperature, particle density, volume from the transverse momentum distributions of charged particles in Pb-Pb and Xe-Xe collisions

2. The application of extensive statistics (Thermal Model) to carry out a systematic analysis of the dependence of chemical freeze-out temperature T_{ch}, strangeness saturation factor γ_s and radius on the charged particle multiplicity $dN_{ch}/d\eta$ in $p - p$, $p-$Pb and Pb-Pb collisions.

1.6   book scope

This book is organized into four parts. Part I introduces the field of high energy physics, study objectives and the scope of this book. Part II is the application of the Tsallis distribution to $p - p$ and Pb-Pb collisions. Part III is the applications of Thermal Model to $p - p$, $p-$Pb and Pb-Pb collisions. Lastly, we present the conclusions, an appendix with the basic kinematics variables and references.

Part II

Applications of non-extensive statistics

Chapter 2

Basic literature

This chapter introduces the non-extensive, Tsallis distribution and its use in the study of transverse momentum p_T distributions of particles produced in high energy collisions. The sections on model description and computing χ^2 values were presented before in [2], while the section on determining thermodynamic parameters has been presented in [1].

2.1   The Tsallis distribution

The Tsallis distribution [17] was first proposed more than three decades ago as a generalization of the Boltzmann-Gibbs distribution and is characterized by three parameters namely, the Tsallis parameter q, the temperature T and the volume V (or the integrated yield dN/dy). Over the years, a vast amount of data has been generated by several collaborations. making use of the various nuclear collisions laboratories around the world and this provides us with an opportunity to test the principles of relativistic thermodynamics and hydrodynamics.

A few years ago it was suggested that the parameters appearing in the Tsallis distribution are the same for a wide range of identified hadrons [18, 19] at $\sqrt{s}$ = 900 GeV in $p - p$ collisions. Subsequently several analyses have appeared which do not support this conclusion; due to the size of the errors and the uncertainties on some of the parameters it is not completely possible to eliminate this possibility at $\sqrt{s}$ = 900 GeV. Further analyses at higher energies are not in support of the original ansatz and led to the proposal of having sequential freeze-outs depending on the particle type [20, 21, 22, 23].

It is well accepted that the transverse momentum distributions in high energy $p - p$ collisions are described by a power law distribution at the Relativistic Heavy Ion Collider (RHIC) [24, 25, 26] as well as at the Large Hadron

Collider [27, 28, 29, 30, 31]. At kinetic freeze-out, we make use of the non-extensive Tsallis distribution, which, being a polynomial one characterized by the Tsallis parameter q, contains information about collective flow.

2.2   The Tsallis non-extensive entropy

In 1988, Tsallis [17] postulated the generalised form of entropy S_q given by

$$S_q \equiv k\frac{1 - \sum_{i=1}^{W} p_i^q}{q - 1},$$

(2.1)

here q is a positive real number, p_i's are the probabilities of finding a system in microstate i, with i running from 1 to W and k is the Boltzmann's constant. The main difference between this distribution and the standard Boltzmann-Gibbs formulation is found in the non-additivity of entropy [32]: taking two independent subsystems A, B, such that the probability of $A + B$ is factorized into $p_{A+B} = p_A p_B$, it should follow that the global entropy becomes

$$S_q(A + B) = S_q(A) + S_q(B) + (1 - q)S_q(A)S_q(B).$$

(2.2)

If one considers the microcanonical ensemble scenario, where all microstates have an exactly fixed energy, volume and particle number, together with the conditions that $p_i = 1/W$ and $\sum_{i=1}^{W} p_i^q = W^{1-q}$, the Tsallis entropy eq. (2.1) now reduces to [17, 32]

$$S_q = k\frac{W^{1-q} - 1}{1 - q} = k\frac{e^{(1-q)\ln W} - 1}{1 - q}.$$

(2.3)

Tsallis applied the $q-$logarithm ($\ln_q x$) which is defined following from the close resemblance of eq. (2.3) to the Boltzmann's entropy $S = k_B \ln W$ as:

$$\ln_q x \equiv \frac{x^{1-q} - 1}{1 - q},$$

(2.4)

whose inverse, is known as the $q-$exponential and is given by [33]

$$e_q^x = [1 + (1 - q)x]^{1/(1-q)}.$$

(2.5)

In the limit, as $q \to 1$, it is straightforward to recover the natural logarithm from the $q-$logarithm. The $q-$logarithm function shall be adopted in the definition of the transverse momentum distribution of particles.

The Tsallis thermodynamic quantities, for a system of massive particles, can be written as integrals over certain combinations of the Tsallis distribution f, defined for any $q \geq 1$ by

$$f(E, q, T, \mu) \equiv \left[1 + (q - 1)\, \frac{E - \mu}{T} \right]^{-\frac{1}{q-1}}, \tag{2.6}$$

which, aside from q and T contains also the chemical potential μ as a parameter. It is obvious that the above reduces to the Boltzmann-Gibbs exponential distribution, in the limit $q \to 1$. The formulation we considered here is thermodynamically consistent, that is, the 1st laws of thermodynamics are satisfied:

$$d\epsilon = T\, ds + \mu\, dn, \tag{2.7}$$

$$dP = s\, dT + n d\mu, \tag{2.8}$$

where ϵ is the energy density, T is the temperature, s is the entropy density, P is the pressure, μ is the chemical potential and n is the particle density. It was shown (see [18, 19, 34] for more details) that n, ϵ, s and P given by the following relations:

$$n \equiv g \int \frac{d^3 p}{(2\pi)^3} f^q, \tag{2.9}$$

$$\epsilon \equiv g \int \frac{d^3 p}{(2\pi)^3} E f^q, \tag{2.10}$$

$$s \equiv -g \int \frac{d^3 p}{(2\pi)^3} \left[f^q \ln_q f - f \right], \tag{2.11}$$

$$P \equiv g \int \frac{d^3 p}{(2\pi)^3} \frac{p^2}{3\, E} f^q. \tag{2.12}$$

Here, we note that the lower case letters ($n, s,$ and ϵ) stand for the corresponding densities, and g is the degeneracy factor which represents the number of possible states (or the number of different combinations of quantum numbers) with the same energy E; all satisfy the thermodynamic consistency conditions:

$$n = \left. \frac{\partial P}{\partial \mu} \right|_T, \quad s = \left. \frac{\partial P}{\partial T} \right|_\mu, \quad T = \left. \frac{\partial \epsilon}{\partial s} \right|_n, \quad \mu = \left. \frac{\partial \epsilon}{\partial n} \right|_s. \tag{2.13}$$

The parameter T is now considered as a temperature

$$T = \left. \frac{\partial E}{\partial S} \right|_{V,N} ,$$

(2.14)

where the entropy S is the Tsallis entropy. The non-extensive parameter q is interpreted as the fluctuations from the Boltzmann distribution as presented in [35]. For completeness, we present the proof in Appendix B.

The main motivation for using the Tsallis distribution over the Boltzmann distribution is that the Boltzmann distribution captures the low p_T range is the transverse momentum spectra, yet the data follows a power law.

2.3   Model description

The Tsallis form of the transverse momentum distribution has been widely used in the analysis of transverse momentum spectra. The reader of this book is encouraged to see [36, 37, 38, 39, 40, 41, 42, 43, 44, 45, 46, 47, 48, 49, 50, 51, 52, 53, 54, 55, 56, 57, 58]. The transverse momentum distribution is given by

$$E \frac{d^3 N}{d^3 p} = gVE \frac{1}{(2\pi)^3} \left[1 + (q-1) \frac{E - \mu}{T} \right]^{-\frac{q}{q-1}} ,$$

(2.15)

where V is the volume. Now, let's introduce rapidity (y), transverse mass $m_T = \sqrt{p_T^2 + m^2}$, energy $E = m_T \cosh y$,

$$\frac{d^2 N}{dp_T dy} = gV \frac{p_T m_T}{(2\pi)^2} \left[1 + (q-1) \frac{m_T \cosh y - \mu}{T} \right]^{-\frac{q}{q-1}} .$$

(2.16)

Following a proposal by [47] who argued that the variables T, V, q, μ in eq. (2.16) have a redundancy for $\mu \neq 0$: to observe this, one can write eq. (2.16) and comparing this to the same equation being considered for finite values of μ (i.e $\mu = 0$ in which one extracts T_0, V_0), we obtain [47]

$$T_0 = T \left[1 - (q-1) \frac{\mu}{T} \right], \qquad \mu \leq \frac{T}{q-1} ,$$

(2.17)

$$V_0 = V \left[1 - (q-1) \frac{\mu}{T} \right]^{\frac{q}{1-q}} ,$$

(2.18)

which implies that the chemical potential may be calculated once the parameters T_0, V_0 and q are known or that T_0 can also be calculated from eq. (2.17). We will use T_0 and V_0 in most of the fits and only come back to $\mu \neq 0$ in chapter 4.

Starting from $m_T = \sqrt{p_T^2 + m^2}$, we can write

$$\begin{aligned}
m_T^2 &= p_T^2 + m^2 \\
m_T dm_T &= p_T dp_T,
\end{aligned} \tag{2.19}$$

this ensures equality of the corresponding invariant yields [59]. Substituting eq. (2.19) into eq. (2.16) gives

$$\frac{1}{m_T}\frac{d^2N}{dm_T dy} = gV\frac{m_T}{(2\pi)^2}\left[1 + (q-1)\frac{m_T \cosh y - \mu}{T}\right]^{-\frac{q}{q-1}}. \tag{2.20}$$

The above equation is a function of m_T only, this implies that for all particle species, a plot of the integrated yield vs. m_T should result in the same slope for different particle species, this is called m_T scaling, see e.g. [60]. It is worth mentioning that m_T scaling is not observed experimentally: we can only determine this if one compare two different particles.

One way of checking that the fit is consistent is by comparing the resulting fit parameter values to some reported parameter values. The extracted parameter V_0 or radius is not quoted in published data, hence one can express the volume in terms of the more easily accessible dN/dy [61], which is found by integrating eq. (2.16)

$$\begin{aligned}
\frac{dN}{dy}\bigg|_{y=0} &= \frac{gV_0}{(2\pi)^2}\int_0^\infty p_T dp_T m_T \left[1 + (q-1)\frac{m_T}{T_0}\right]^{-\frac{q}{q-1}} \\
&= \frac{gVT_0}{2\pi)^2}\left[\frac{(2-q)m^2 + 2mT_0 + 2T_0^2}{(2-q)(3-2q)}\right] \times \\
&\quad \left[1 + (q-1)\frac{m}{T_0}\right]^{-\frac{1}{q-1}}, \tag{2.21}
\end{aligned}$$

where m is the mass of the respective particle. Notice that in eq. (2.21), we have T_0, this is because the integral is performed at $\mu = 0$. Now, we express V_0 in terms of dN/dy, T_0 and q from the above and substituting in eq. (2.16) result in [61]

$$\frac{1}{p_T}\frac{d^2N}{dp_Tdy}\bigg|_{y=0} = \frac{m_T}{T_0}\frac{dN}{dy}\bigg|_{y=0}\frac{(2-q)(3-2q)}{(2-q)m^2+2mT_0+2T_0^2}$$
$$\left[1+(q-1)\frac{m}{T_0}\right]^{\frac{1}{q-1}}\left[1+(q-1)\frac{m_T}{T_0}\right]^{-\frac{q}{q-1}}. \tag{2.22}$$

For identified charged particles produced in high energy collisions, eq. (2.16) is given by a sum over the most abundant charged particle species e.g π^+, K^+, p [49] as

$$\frac{1}{p_T}\frac{d^2N_{ch}}{dp_Tdy}\bigg|_{y=0} \cong \sum_{\pi^+,K^+,p^+} g_i V_0 \frac{m_{T,i}}{(2\pi)^2}\left[1+(q-1)\frac{m_{T,i}}{T_0}\right]^{-\frac{q}{q-1}}, \tag{2.23}$$

with degeneracy $g_{\pi^+} = 1$, $g_{K^+} = 1$ and $g_p = 2$. If we combine the particles and antiparticles together (since they carry equal masses), and let $i = \pi^+, K^+, p$, we can also write eq. (2.23) as

$$\frac{1}{p_T}\frac{d^2N_{ch}}{dp_Tdy}\bigg|_{y=0} \cong \frac{2V_0}{(2\pi)^2}\sum_{i=1}^{3} g_i m_{T,i}\left[1+(q-1)\frac{m_{T,i}}{T_0}\right]^{-\frac{q}{q-1}}. \tag{2.24}$$

The factor 2 appearing on the right hand side of eq. (2.23) takes into account the contributions from antiparticles π^-, K^- and $\bar{p}$. For the summation $\sum_{i=1}^{3}$, $i = 1$ for pions, 2 for kaons and 3 for protons. For charged particles, the data are mostly given in terms of pseudorapidity, and using the result from Appendix I, eq. (A.8)

$$\frac{dN}{dp_Td\eta} = \frac{p_T}{m_T}\frac{dN}{dp_Tdy}, \tag{2.25}$$

which introduces an extra factor of p_T/m_T, one has to take care of this factor when fitting the transverse momentum spectra data with the Tsallis distribution.

The advantage of using eq. (2.22) over eq. (2.16) is that one can estimate the value of dN/dy which is a known quantity and this allows for comparison and a further validation of the obtained fit parameters.

2.3.1 Determining thermodynamic parameters

In heavy-ion collisions at the LHC, hadronic matter is created at a very high energy density. After the initial very hot stage the system expands, reaches chemical equilibrium and then finally freezes out in a stage usually referred to as the kinetic freeze-out stage. The Tsallis thermodynamic parameters: energy density, pressure, entropy density, temperature, together with particle number density for a system of charged particles at kinetic freeze-out are determined as integrals of the Tsallis distribution e.q (2.6) as given in [18, 19, 34].

We integrate eq. (2.9) through to eq. (2.12) and write out the simplified forms of these equations in the itemized list below:

- Energy density at kinetic freeze-out

$$\epsilon \cong 2 \sum_{i=1}^{3} g_i \int \frac{d^3p}{(2\pi)^3} E_i \left[1 + (q-1) \frac{E_i}{T_0} \right]^{-\frac{q}{q-1}} , \qquad (2.26)$$

the leading factor 2 accounts for the corresponding antiparticles. Taking an integral over angular variables, one obtains

$$\epsilon \cong \sum_{i=1}^{3} g_i \int \frac{p^2 dp}{\pi^2} E_i \left[1 + (q-1) \frac{E_i}{T_0} \right]^{-\frac{q}{q-1}} . \qquad (2.27)$$

- Pressure density at kinetic freeze-out: the pressure plays an important role in the hydrodynamic description of heavy-ion collisions, e.g. in the study of shock waves or the speed of sound in a hadronic gas. This is given by

$$p \cong \sum_{i=1}^{3} g_i \int \frac{p^4 dp}{\pi^2} \frac{1}{3E_i} \left[1 + (q-1) \frac{E_i}{T_0} \right]^{-\frac{q}{q-1}} . \qquad (2.28)$$

- Particle number density at kinetic freeze-out similarly

$$n \cong \sum_{i=1}^{3} g_i \int \frac{p^2 dp}{\pi^2} \left[1 + (q-1) \frac{E_i}{T_0} \right]^{-\frac{q}{q-1}} . \qquad (2.29)$$

- Entropy density at kinetic freeze-out: the entropy is an important quantity because it plays a major role in hydrodynamic expansion calculations where entropy is sometimes assumed to be conserved when going from the quark-gluon plasma phase to the hadronic phase. This is for example the case in the Bjorken model [62] and other model calculations. It is difficult to relate it directly to a measurable quantity and it is often indirectly linked to the particle number. The entropy density can also be calculated as follows:

$$s \cong -2 \sum_{i=1}^{3} g_i \int \frac{d^3p}{(2\pi)^3} \left[f^q \ln_q f - f \right] , \qquad (2.30)$$

solving for $\ln_q f$

$$
\begin{aligned}
\ln_q f &= \frac{f^{1-q}-1}{q-1}; \quad f > 0, \ q \neq 1 \\[2mm]
&= \frac{\left[1+(q-1)\frac{E}{T_0}\right]^{-\frac{(1-q)}{(q-1)}} - 1}{1-q} \\[2mm]
&= -\frac{E}{T_0}
\end{aligned}
\tag{2.31}
$$

now, substituting for f and $\ln_q f$ in e.q. (2.30) and simplifying results in

$$
s \cong \sum_{i=1}^{3} g_i \int \frac{p^2 dp}{\pi^2} \left\{ \frac{E_i}{T_0} \left[1+(q-1)\frac{E_i}{T_0}\right]^{-\frac{q}{q-1}} + \left[1+(q-1)\frac{E_i}{T_0}\right]^{-\frac{1}{q-1}} \right\}.
\tag{2.32}
$$

The above equations are used to calculate the ϵ, p, s, and n for a system of charged particles.

2.3.2 Average momentum

The average momentum $\langle p_T \rangle$ data for emitted particles are readily available from published articles in most cases; thus, for us to check the integrity of our formulation, we calculate $\langle p_T \rangle$ following the definition in [63] as

$$
\langle p_T \rangle = \frac{\int_0^\infty dp_T \, p_T \frac{dN}{dp_T dy}\Big|_{y=0}}{\int_0^\infty dp_T \frac{dN}{dp_T dy}\Big|_{y=0}},
\tag{2.33}
$$

and compare our results to the published values.

2.3.3 Computing χ^2 values

A chi-squared χ^2 test, is a statistical test that compares the actual data to a modelled result. The χ^2 values are determined using values of the transverse momentum distribution for a given value of p_T. Let the theoretical result $N_i(T, q, dN/dy)$ be given by

$$
N_i(T, q, dN/dy) \equiv \frac{d^2 N}{dp_T dy}\Big|_{p_T = p_T^i}
\tag{2.34}
$$

for the data at $\sqrt{s} = 900\,\text{GeV}$ and by

$$N_i(T,q,dN/dy) \equiv \frac{1}{2\pi p_T} \left. \frac{d^2 N}{d p_T dy}\right|_{p_T = p_T^i}. \tag{2.35}$$

for all other beam energies: the difference is resulting from the $\frac{1}{2\pi p_T}$ factor present in the published experimental data. Both of these are calculated using the Tsallis distribution with the optimised values of q, T and dN/dy. This theoretical result is then compared with the experimental values to determine a value of χ^2

$$\chi_y^2 = \sum_{\forall i} \left(\frac{N_i(T,q,dN/dy) - E_i}{\sigma_i} \right)^2, \tag{2.36}$$

where E_i is the experimental value of the momentum distribution at the same value of p_T and σ_i is the experimental error on the distribution: here we have added the index y to χ^2 to emphasize that these values refer to the error bars on the ordinate (y) axis.

2.4   Summary

In this chapter, we presented the non-extensive statistics and their applications in the study of transverse momentum p_T distributions. The literature presented in this chapter will inform the data analysis in the next chapter.

Chapter 3

Tsallis fits to $p - p$ collisions without radial flow

This chapter presents the application of Tsallis non-extensive statistics in the analysis of p_T distributions in $p - p$ collisions. The Tsallis distribution with zero chemical potential is made use of. We use published data from the ALICE and CMS experiments to extract values of the fit parameters (q, T and dN/dy). The results from the analysis of ALICE data were presented in [2].

3.1   Applications of non-extensive statistics

In this study, we investigate in detail one particular form of power law distribution which has been used extensively in the literature [36, 37, 38, 39, 40, 41, 42, 43, 44, 45, 46, 47, 48, 49, 50, 51, 52, 53] given by [17]:

$$f(E) \equiv \left[1 + (q - 1)\frac{E}{T} \right]^{-1/(q-1)} , \tag{3.1}$$

where q and T are two parameters to be determined. It is referred to as the Tsallis distribution [17] and forms the basis for non-extensive statistical thermodynamics. This is accomplished through:

1. **Applications of Tsallis distribution to $p - p$ collisions.**

   We seek to determine the parameters appearing in the Tsallis distribution as precisely as possible at beam energies of $\sqrt{s} = 0.9, 2.76, 7$ and 13 TeV. We make use of the ALICE and CMS data in this analysis. The results obtained in the analysis of ALICE data are published in [2].

2. **Applications of Tsallis distribution to Pb-Pb and Xe-Xe collisions.**

   We determine thermodynamic parameters: energy density, pressure, entropy density, temperature, particle density, volume from the transverse momentum distributions of charged particles in Pb-Pb collisions at 2.76 and 5.02 TeV as well as in Xe-Xe collisions at $\sqrt{s_{NN}} = 5.55$ TeV as a function of centrality. The results obtained in the analysis of Pb-Pb collisions are published in [1].

3.2 Method

We make use of a ROOT [64] macro, which utilises TMINUIT [64] in the computation of the χ^2 values via a minimisation procedure. The reader is refereed to the TMINUIT package for more details on the fit procedure. For our purposes, it is sufficient to mention that, TMINUIT will make use of the initial fit parameter values to compute the best estimates of the fit parameters. It returns the final fit parameters that gives the minimum χ^2.

In using the developed macro, a visual inspection of the fit and an analysis of the fit/data ratio provides an indication of how good the model describes experimental data.

- define fit function and other functions

- define canvas for displaying results

- input data

- initialise fit parameters

- fit the data with the fit function

- perform other required calculations

- display results and write out results to files

Presented above is a schematic showing the outline of a ROOT macro used in extracting fit parameters: in defining the function, we ensure that all parameters are clearly defined.

The input data should correspond to the respective particle. It is not trivial to obtain a set of values for input parameters: as a guide, one has to think about what size the radius of the system should be, what value of kinetic freeze-out temperature can one expect and what are the lower and upper bounds for the q parameter.

3.3 Fits to p_T spectra at $\sqrt{s} = 900$ GeV with ALICE

The Tsallis parameters were extracted by fitting the experimental results published by the ALICE collaboration [65] to eq. (2.22): the data at $\sqrt{s} = 900$ GeV has the smallest range in p_T (for all the ALICE collaboration results considered in this book), of about an order of magnitude less than the experimental data at $\sqrt{s} = 2.76$ GeV and $\sqrt{s} = 7$ TeV with ALICE. The resulting values for the beam energy of $\sqrt{s} = 900$ GeV are shown in Table 3.1.

TABLE 3.1: Fit results at $\sqrt{s} = 900$ GeV using data from the ALICE [65] collaboration. Note the very large errors on the values of T_0 for protons and antiprotons.

Particle	dN/dy	q	T_0 (MeV)	χ^2 / NDF	$\langle p_T \rangle$ (GeV)
π^+	1.488 ± 0.019	1.148 ± 0.005	70 ± 2	$22.73/30$	0.408 ± 0.023
π^-	1.480 ± 0.018	1.145 ± 0.005	72 ± 2	$15.83/30$	0.408 ± 0.022
K^+	0.184 ± 0.004	1.175 ± 0.016	57 ± 12	$13.02/24$	0.663 ± 0.142
K^-	0.182 ± 0.004	1.161 ± 0.015	64 ± 12	$6.214/24$	0.641 ± 0.202
p	0.083 ± 0.002	1.16 ± 0.025	17 ± 28	$14.52/21$	0.773 ± 0.270
$\bar{p}$	0.079 ± 0.002	1.132 ± 0.025	52 ± 30	$13.82/21$	0.766 ± 0.250

The fit results presented in Table 3.1 show some very large errors on the values of the temperature T_0 for protons and anti-protons: the central values for T_0 differ by a factor of three between protons and anti-protons. We shall explore this further in an attempt to find an explanation for this disparity. Apart from large errors, the central values for T_0 deviate clearly from those obtained for $\pi^\pm$ and $K^\pm$.

An earlier analysis at the same beam energy reported consistent values for T_0, however with large error bars [18]. A comparison is shown in Figure 3.1 in which the central values of current result are not in fully agreement with [18].

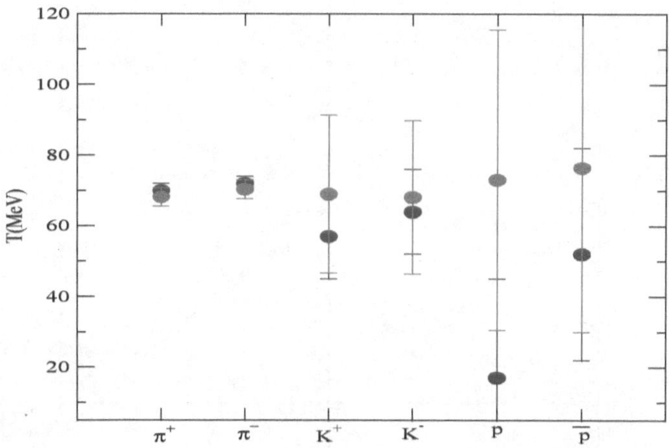

FIGURE 3.1: A comparison of the values of temperature for different hadron species for $p - p$ collisions $\sqrt{s} = 900$ with the ALICE [65] collaboration. The fit result reported by [18] is represented by red circles while the results presented in this book are represented by blue circles.

The transverse momentum distributions at beam energy of $\sqrt{s} = 900\,\text{GeV}$ are shown in Figure 3.2 where it can be seen that both the protons and the antiprotons fits describe the data very well.

However, the resulting parameters have large errors and are quite different: T_0 for protons is 17 ± 29 MeV with q being 1.158 ± 0.028; for antiprotons the value of T_0 is 52 ± 33 MeV while q is 1.132 ± 0.028. The central values of T_0 thus differ by a factor of 3 between protons and anti-protons. However, these results are consistent within error bars as shown in Figure 3.3.

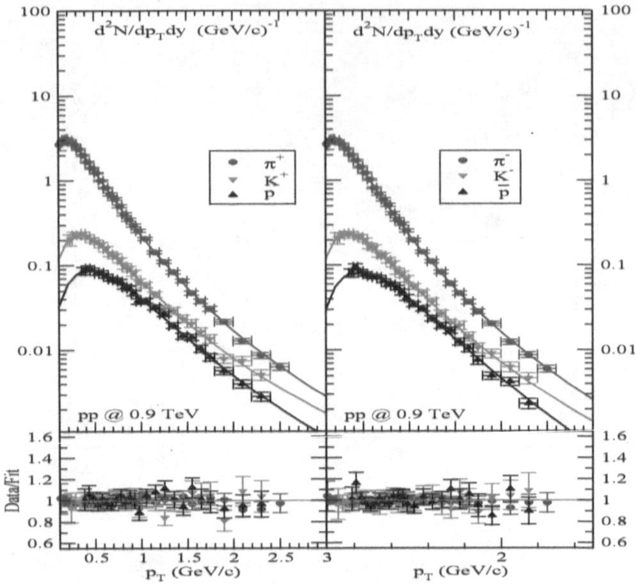

FIGURE 3.2: Fits to the transverse momentum distributions, using the Tsallis distribution eq. (2.22), of π^+ (blue) K^+ (red) and protons (black) at 900 GeV [65] in the left panel. In the right panel π^- (blue), K^- (red), antiprotons (black) as measured by the ALICE collaboration at 900 GeV.

The difference in extracted values of T_0 comes as a surprise since the integrated proton and antiproton yields are equal, within the uncertainties: it is natural to expect slight differences.

Our interpretation is that within the measured range of transverse momenta, it is not possible to determine precisely the values of the Tsallis parameters. To emphasize this we show in Figure 3.3 the contour of $1 - \sigma$ uncertainties in the $T_0 - q$ plane.

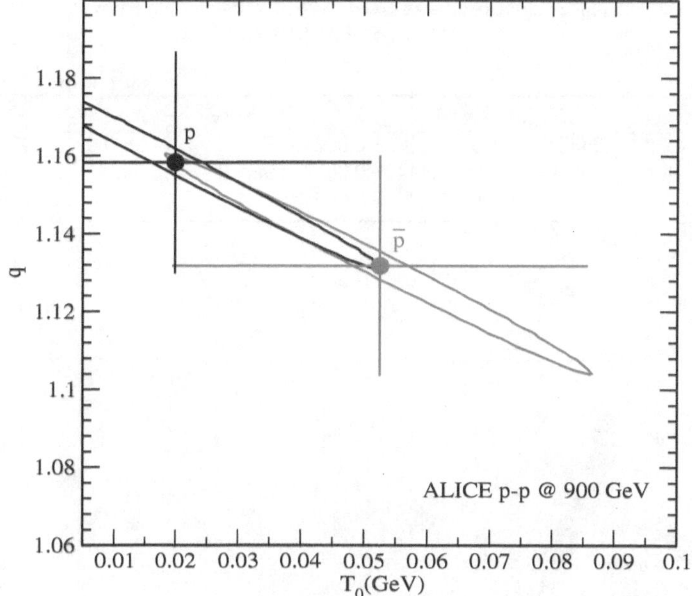

FIGURE 3.3: Contours in the $T_0 - q$ plane showing lines with 1 standard deviation from the minimum χ^2 values as well as the values of the parameters q and T_0 (corresponding to the minimum χ^2) with error bars. Values for protons are in black while those for antiprotons are given in red. Note the large difference of more than a factor of 3 between the central temperature values.

The contour plot in Figure 3.3 suggests that despite such a difference in the central values of T_0 for the protons and antiproton, these central values agree within 1 standard deviation from the minimum χ^2 values.

3.4 Fits to p_T spectra at $\sqrt{s} = 2.76$ TeV with ALICE

The transverse momentum spectra at $\sqrt{s} = 2.76$ TeV have been measured in a range extending up to about 20 GeV/c. The fit is shown in Figure 3.4 and the resulting values of the Tsallis parameters are presented in Table 3.2 below.

It can be seen that the values of q and T_0 are much more constrained than in the previous case for $\sqrt{s} = 900$ GeV. If the possibility of common values could be completely discarded in the previous case, there is no doubt here that the values are different.

Particle	dN/dy	q	T_0 (MeV)	χ^2 / NDF	$\langle p_T \rangle$ (GeV)
$\pi^+ + \pi^-$	3.967 ± 0.083	1.149 ± 0.002	77 ± 1	$242.8/60$	0.442 ± 0.011
$K^+ + K^-$	0.463 ± 0.010	1.144 ± 0.002	96 ± 3	$10.55/55$	0.408 ± 0.024
$p + \bar{p}$	0.209 ± 0.006	1.121 ± 0.005	87 ± 9	$26.35/46$	0.408 ± 0.056

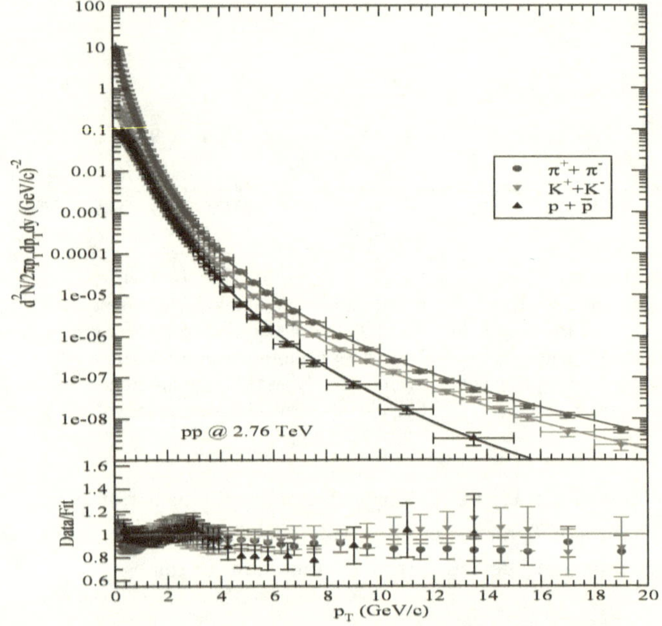

FIGURE 3.4: Fits to the transverse momentum distributions, using the Tsallis distribution eq. (2.22), $\pi^+ + \pi^-$ (blue), $K^+ + K^-$ (red), protons and antiprotons (black) as measured by the ALICE collaboration at 2.76 TeV [66].

The transverse momentum distributions are shown in Figure 3.4, where it can be seen that the fits to both the protons and the antiprotons describe the data very well. Also on the bottom panel in Figure 3.4 , the ratio of the Data/Fit shows an excellent agreement over the entire p_T range.

3.5 Fits to p_T spectra at $\sqrt{s} = 5.02$ TeV with ALICE

The transverse momentum spectra at $\sqrt{s} = 5.02$ TeV we have used the interpolated data as presented by the ALICE collaboration [67] (in a range extending up to about 20 GeV/c) by interpolating between 2.76 TeV and 7 TeV. The fit is shown in Figure 3.5 and the resulting values of the Tsallis parameters are presented in Table 3.3 below.

TABLE 3.3: Fit results at $\sqrt{s} = 5.02$ TeV, using the interpolated data as given by the ALICE collaboration [67].

Particle	dN/dy	q	T_0 (MeV)	χ^2 / NDF	$\langle p_T \rangle$ (GeV)
$\pi^+ + \pi^-$	4.452 ± 0.095	1.155 ± 0.002	76 ± 2	$266.3/60$	0.452 ± 0.016
$K^+ + K^-$	0.530 ± 0.011	1.150 ± 0.005	99 ± 6	$12.11/55$	0.750 ± 0.049
$p + \bar{p}$	0.235 ± 0.007	1.126 ± 0.005	91 ± 9	$18.89/46$	0.877 ± 0.059

Again, due to the quality of the data, the values of q and T_0 are much more constrained than for $\sqrt{s} = 900$ GeV. The resulting values at $\sqrt{s} = 5.02$ TeV are shown in Table 3.3.

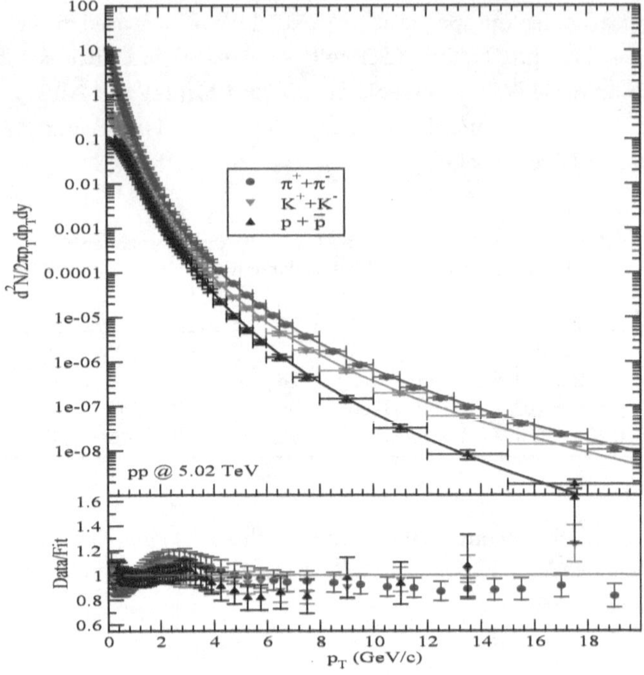

FIGURE 3.5: Fits to the transverse momentum distributions, using the Tsallis formula given in eq. (2.22), of $\pi^+ + \pi^-$ (blue), $K^+ + K^-$ (red), protons and antiprotons (black) as given by the ALICE collaboration at 5.02 TeV [67] by interpolating between the measured 7 TeV [68] data and the measured 2.76 TeV data.

The transverse momentum distributions shown in Figure 3.5 on the top panel: where it can be seen that the fits to all protons, kaons and protons describe the data very well. Also on the bottom panel, the ratio of the Data/Fit shows an excellent agreement over the entire p_T range.

3.6 Fits to p_T spectra at $\sqrt{s} = 7$ TeV with ALICE

The transverse momentum spectra at $\sqrt{s} = 7$ TeV in $p - p$ collisions have also been measured [68] in a range extending up to about 20 GeV/c. The fit is shown in Figure 3.6 and the resulting values of the Tsallis parameters are presented in Table 3.4 below.

TABLE 3.4: Fit results at $\sqrt{s} = 7$ TeV, using data from the ALICE collaboration [68].

Particle	dN/dy	q	T_0 (MeV)	χ^2/NDF	$\langle p_T \rangle$ (GeV)
$\pi^+ + \pi^-$	4.778 ± 0.101	1.158 ± 0.002	76 ± 2	$331.7/55$	0.460 ± 0.017
$K^+ + K^-$	0.573 ± 0.011	1.155 ± 0.005	100 ± 6	$27.54/48$	0.777 ± 0.052
$p + \bar{p}$	0.251 ± 0.007	1.129 ± 0.005	94 ± 9	$20.26/46$	0.903 ± 0.061

The resulting values for dN/dy at mid-rapidity are fully compatible (if not identical) to the values quoted by the ALICE collaboration [68] using a slightly different parametrization for the transverse momentum distribution.

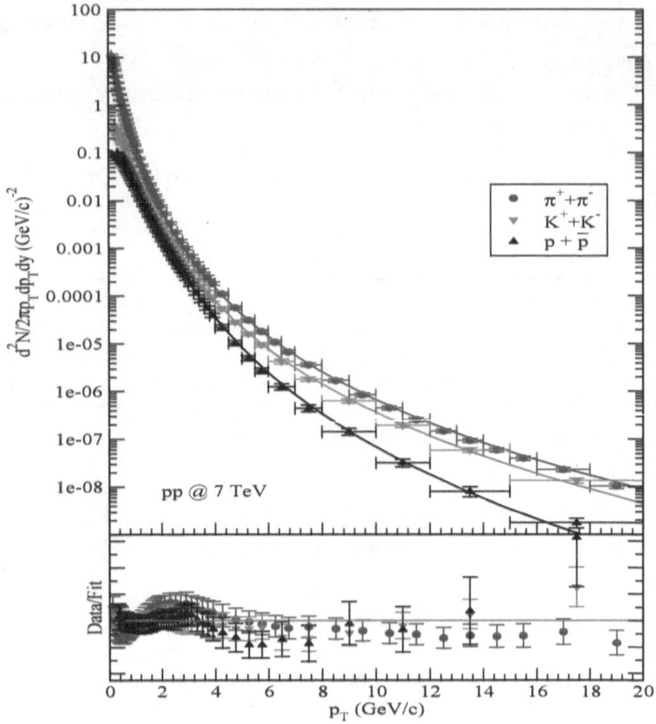

FIGURE 3.6:   Fits to the transverse momentum distributions, using the Tsallis formula given in eq. (2.22), of $\pi^+ + \pi^-$ (blue), $K^+ + K^-$ (red), protons and antiprotons (black) as given by the ALICE collaboration 7 TeV [68].

The transverse momentum distributions shown in Figure 3.6 on the top panel: where it can be seen that the fits to all protons, kaons and protons describe the data very well. Also on the bottom panel, the ratio of the Data/Fit shows an excellent agreement over the entire p_T range.

3.7 Fits to p_T spectra at $\sqrt{s} = 900$ GeV with CMS

The Tsallis parameters were extracted by fitting the experimental results published by the CMS Collaboration [69, 28, 29, 70, 71] to eq. (2.22). The experimental results are in the range extending up to 2 GeV in p_T. This is the same range in p_T as the p_T spectra as measured by the ALICE collaboration at a beam energy of $\sqrt{s} = 900$ GeV.

The transverse momentum spectra at $\sqrt{s} = 900$ GeV in $p - p$ collisions have also been measured [28] in a range extending up to about 2 GeV/c. The fit is shown in Figure 3.7 and the resulting values of the Tsallis parameters are presented in Table 3.5 below:

TABLE 3.5: The extracted values of dN/dy, q, T_0 and χ^2/NDF
for $p - p$ collisions at $\sqrt{s} = 900$ GeV with CMS

Particle	dN/dy	q	T_0 (MeV)	χ^2/NDF
π^+	1.919 ± 0.019	1.164 ± 0.005	67 ± 2	$13.04/19$
π^-	1.892 ± 0.019	1.167 ± 0.005	66 ± 2	$19.29/19$
K^+	0.239 ± 0.007	1.158 ± 0.028	78 ± 16	$11.12/14$
K^-	0.241 ± 0.008	1.182 ± 0.029	64 ± 17	$10.24/14$
p	0.109 ± 0.001	1.109 ± 0.001	58 ± 16	$18.58/24$
$\bar{p}$	0.104 ± 0.001	1.147 ± 0.014	47 ± 16	$21.30/24$

In Table 3.5 , the resulting q parameters are quite different for the protons and antiprotons; for protons $q = 1.109 \pm 0.001$; for antiprotons the value is $q = 1.147 \pm 0.014$; The central values of T_0 for protons and anti-protons are in agreement. For the rest of particle and anti-particle pairs, all fit parameters are in agreement.

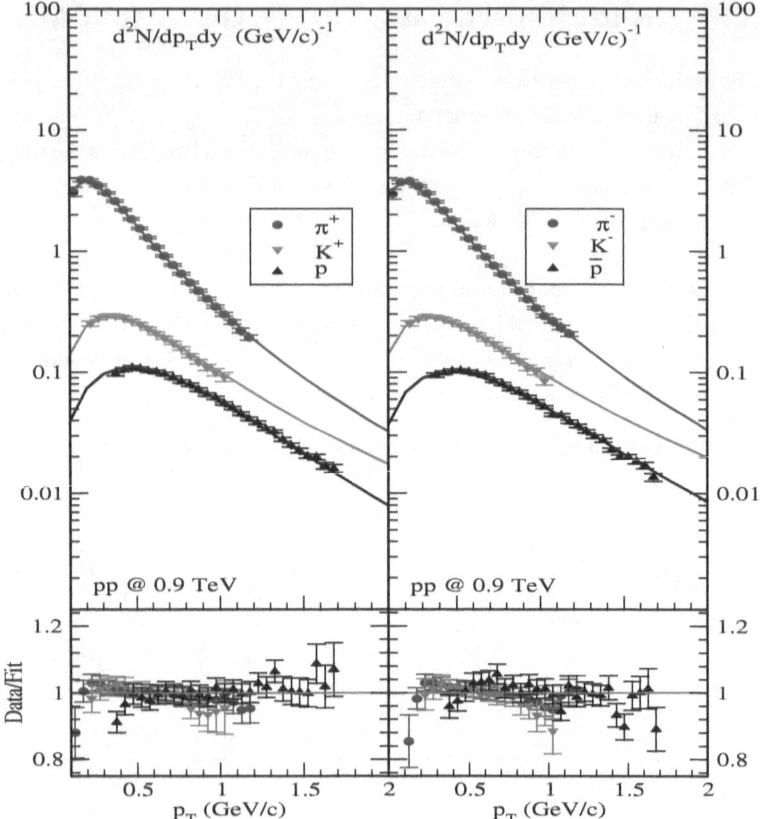

FIGURE 3.7: Fits to the transverse p_T distributions, using the Tsallis distribution eq. (2.22). The CMS data at 900 GeV are as follows: pions are in blue, kaons in red and protons in black. The solid curves represent the fit to data. The Data/Fit ratios are presented below the corresponding p_T distributions.

The transverse momentum distributions at beam energy of $\sqrt{s} = 900\,\text{GeV}$ are shown in Figure 3.7 where it can be seen that for all particles, the fits describe the data very well over the entire p_T range.

3.8 Fits to p_T spectra at $\sqrt{s} = 2.76$ TeV with CMS

The transverse momentum spectra at $\sqrt{s} = 2.36$ TeV in $p-p$ collisions have also been measured [71] in a range extending up to about 2 GeV/c. The fit is shown in Figure 3.8 and the resulting values of the Tsallis parameters are presented in Table 3.6 below:

TABLE 3.6: The extracted values of dN/dy, q, T_0 and χ^2/NDF for $p-p$ collisions at $\sqrt{s} = 2.76$ TeV with CMS [71].

Particle	dN/dy	q	T_0 (MeV)	χ^2/NDF
π^+	2.439 ± 0.025	1.189 ± 0.005	61 ± 2	$14.71/19$
π^-	2.381 ± 0.024	1.184 ± 0.005	63 ± 2	$16.07/19$
K^+	0.311 ± 0.012	1.162 ± 0.031	87 ± 18	$11.45/14$
K^-	0.301 ± 0.011	1.147 ± 0.030	96 ± 18	$16.41/14$
p	0.141 ± 0.002	1.166 ± 0.017	49 ± 18	$27.43/24$
$\bar{p}$	0.133 ± 0.002	1.129 ± 0.015	90 ± 17	$28.41/24$

In Table 3.6 , the resulting values of the q parameter are quite different for the proton antiproton pair: for protons $q = 1.166 \pm 0.017$ while for antiprotons the value is $q = 1.29 \pm 0.015$. The resulting values of the parameter T_0 are also not in agreement for the proton antiproton pair: for protons is $T_0 = 49 \pm 18$ MeV while and for antiprotons the value is $T_0 = 90 \pm 17$ MeV. For the rest of particle and antiparticle pairs, all fit parameters are in agreement within error.

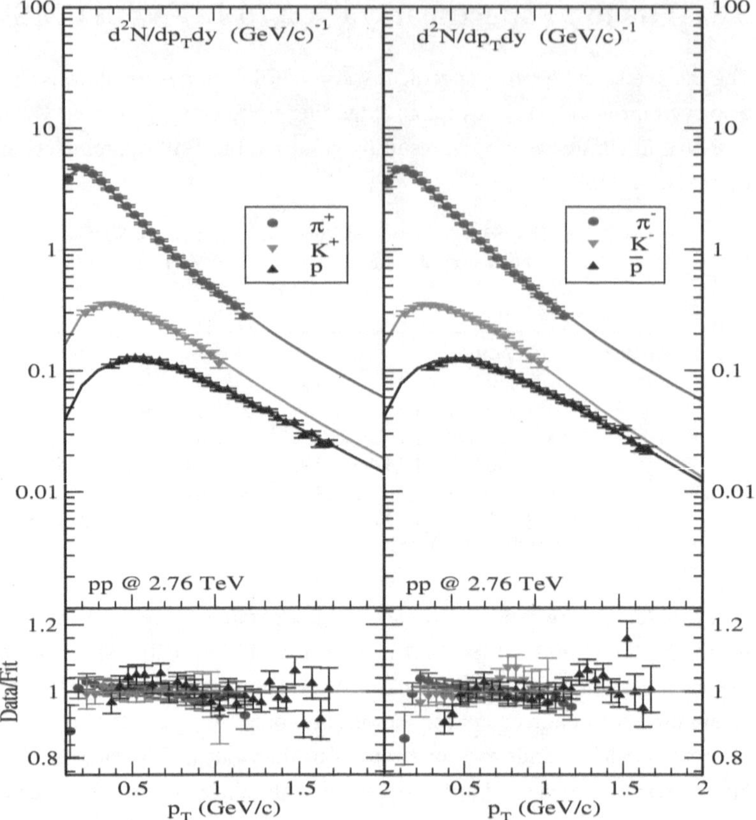

FIGURE 3.8: Fits to the transverse p_T distributions, using the Tsallis distribution eq. (2.22). The CMS data at 2.76 TeV are as follows: pions are in blue, kaons in red and protons in black. The solid curves represent the fit to data. The Data/Fit ratios are presented below the corresponding p_T distributions.

The transverse momentum distributions at beam energy of $\sqrt{s} = 2.76$ TeV are shown in Figure 3.8 where it can be seen that for all particles, the fits describe the data very well over the entire p_T range.

3.9 Fits to p_T spectra at $\sqrt{s} = 7$ TeV with CMS

The transverse momentum spectra at $\sqrt{s} = 7$ TeV in $p - p$ collisions have also been measured [71] in a range extending up to about 2 GeV/c. The fit is shown in Figure 3.9 and the resulting values of the Tsallis parameters are presented in Table 3.7 below:

TABLE 3.7: The extracted values of dN/dy, q, T_0 and χ^2/NDF for $p - p$ collisions at $\sqrt{s} = 7$ TeV with CMS [71].

Particle	dN/dy	q	T_0 (MeV)	χ^2/NDF
π^+	3.053 ± 0.033	1.203 ± 0.006	59 ± 2	$14.29/19$
π^-	3.007 ± 0.032	1.202 ± 0.005	60 ± 2	$11.36/19$
K^+	0.383 ± 0.017	1.152 ± 0.034	102 ± 20	$11.07/14$
K^-	0.397 ± 0.020	1.186 ± 0.036	83 ± 21	$13.30/14$
p	0.180 ± 0.004	1.184 ± 0.021	52 ± 22	$21.22/24$
$\bar{p}$	0.180 ± 0.004	1.19 ± 0.023	45 ± 23	$15.47/24$

In Table 3.7 , all parameters dN/dy, q and T_0 are are in agreement for each particle and antiparticle pairs; however, the temperature values for protons and kaons have huge errors.

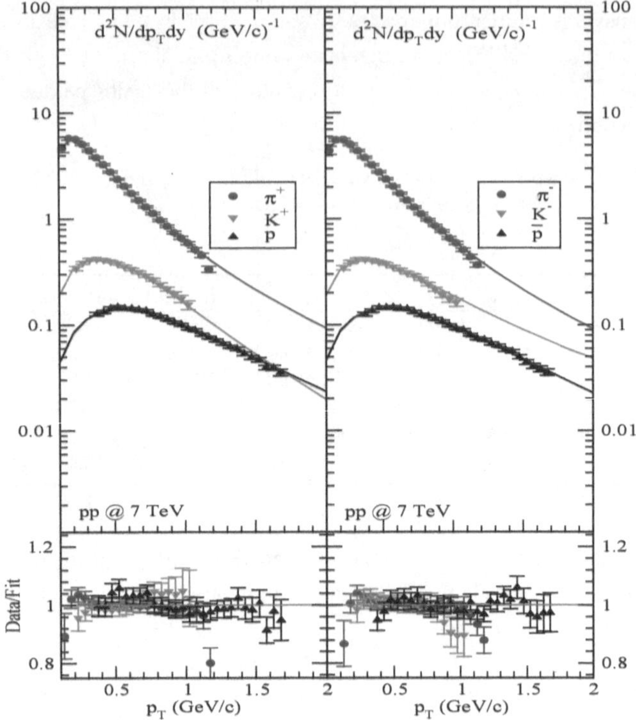

FIGURE 3.9: Fits to the transverse p_T distributions, using the Tsallis distribution eq. (2.22). The CMS data at 7 TeV are as follows: pions are in blue, kaons in red and protons in black. The solid curves represent the fit to data. The Data/Fit ratios are presented below the corresponding p_T distributions.

The transverse momentum distributions at beam energy of $\sqrt{s} = 7$ TeV are shown in Figure 3.9 where it can be seen that for all particles, the fits describe the data very well over the entire p_T range.

3.10 Fits to p_T spectra at $\sqrt{s} = 13$ TeV with CMS

The transverse momentum spectra at $\sqrt{s} = 13$ TeV in $p - p$ collisions have also been measured [70] in a range extending up to about 2 GeV/c. The fit is shown in Figure 3.10 and the resulting values of the Tsallis parameters are presented in Table 3.8 below:

TABLE 3.8: The extracted values of dN/dy, q, T_0 and χ^2/NDF for $p - p$ collisions at $\sqrt{s} = 13$ TeV with CMS [70].

Particle	dN/dy	q	T_0 (MeV)	χ^2/NDF
π^+	2.845 ± 0.038	1.215 ± 0.008	57 ± 3	$12.543/19$
π^-	2.72 ± 0.034	1.191 ± 0.008	67 ± 3	$12.72/19$
K^+	0.335 ± 0.030	1.142 ± 0.069	106 ± 40	$10.828/14$
K^-	0.335 ± 0.033	1.155 ± 0.072	100 ± 42	$10.323/14$
p	0.169 ± 0.007	1.213 ± 0.037	18 ± 37	$17.892/23$
$\bar{p}$	0.162 ± 0.006	1.189 ± 0.036	48 ± 38	$9.383/23$

The fit results presented in Table 3.8 shows some very large errors on the values of the temperature T_0 for protons and anti-protons: the central values for T_0 differ by a factor greater than two and a half between protons and anti-protons. For kaons, there are large errors on T_0 while the central values are comparable.

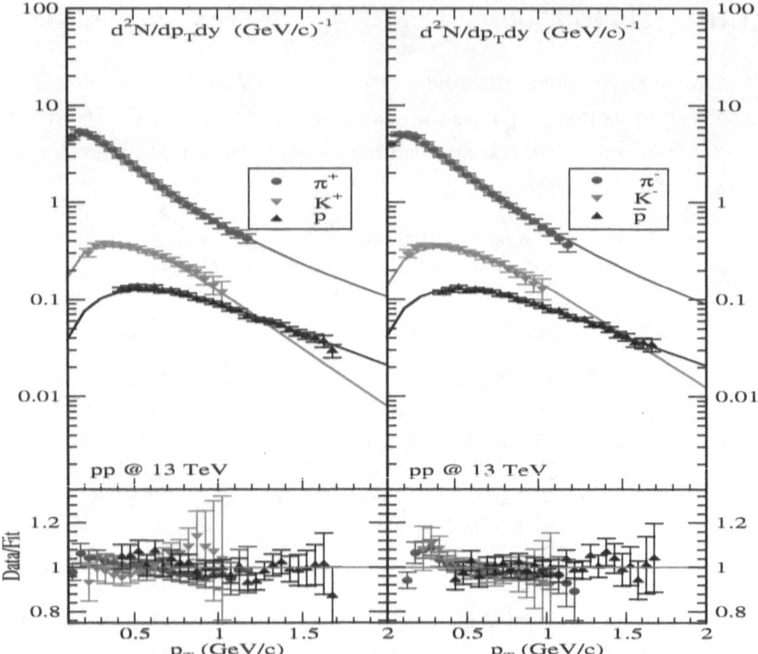

FIGURE 3.10: Fits to the transverse p_T distributions, using the
Tsallis distribution eq. (2.22). The CMS data at 13 TeV are as
follows: pions are in blue, kaons in red and protons in black.
The solid curves represent the fit to data. The Data/Fit ratios
are presented below the corresponding p_T distributions.

The transverse momentum distributions at beam energy of $\sqrt{s} = 13$ TeV
are shown in Figure 3.10 where it can be seen that for all particles, the fits
describe the data very well over the entire p_T range.

3.11   Analysis of results

In order to make sense of the results presented in the previous section: we construct contour plots in the $T_0 - q$ plane with ellipses corresponding to fixed values of deviations from the minimum χ^2 values, the results are presented in Figure 3.11. The 1 standard deviation for the fits to the results of the ALICE collaboration are shown in red; the 2 standard deviation are shown in blue while those with 3 standard deviation are shown in black.

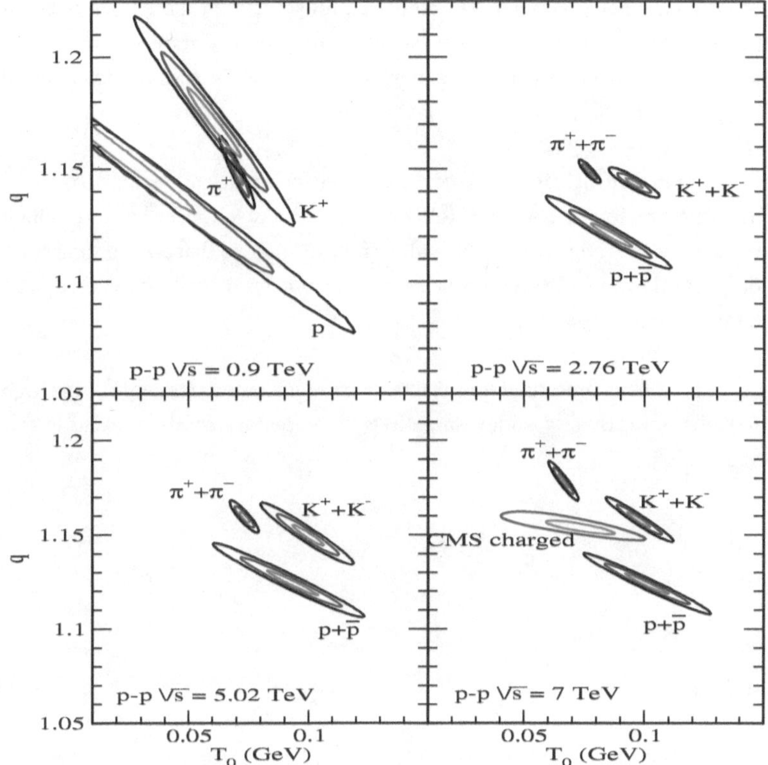

FIGURE 3.11: Contours in The $T_0 - q$ plane showing lines of 1 standard deviation from the minimum χ^2 in red. 2 standard deviations are shown in blue. Those for 3 standard deviation from the minimum χ^2 are shown in black. The upper left panel is for data at $\sqrt{s} = 0.9$ TeV [65]; the upper right panel is for 2.76 TeV [66]. The lower left panel is for 5.02 TeV [67] while the lower right panel is for 7 TeV [68]. For comparison the contours are also shown for results from the CMS collaboration [69] for charged particles at 7 TeV.

We notice that the 7 TeV case has been discussed previously [38, 46], within the framework of the Tsallis distribution. The authors from [38] use a different form of the Tsallis distribution, not having a factor m_T on the right-hand side of eq. (2.22) and hence obtain higher values of the temperature T_0.

Within the framework considered here, the values for q and T_0 obtained from the ALICE collaboration data are consistent with those of the CMS collaboration, as shown in the contour plot presented in Figure. 3.11. The CMS contours are being situated as roughly equidistant from the ALICE ones for pions, kaons and protons. This comes as a surprise since at those large values of the transverse momentum p_T, hard scattering processes are presumed to be dominant.

From Figure 3.11, it can be seen that while the pions and kaons do overlap in a small region at $\sqrt{s} = 0.9$ TeV, this is not the case for the protons, albeit there is a wide range of possible values for the latter so that an eventual complete overlap for all three particles π^+, K^+, p cannot really be excluded in light of the results above.

From the fits presented, the Tsallis distribution simulates some of the high p_T behaviour, thus, a wider range in p_T is necessary to really exclude this eventuality.

The data from CMS collaboration at all energies are in the same p_T range from about $0.1 - 2$ GeV while those from the ALICE collaboration extends to about 20 GeV in p_T for the $2.76, 5.02$ and 7 TeV experiments. The resulting contours in the $T_0 - q$ plane are presented in Figure 3.12 below:

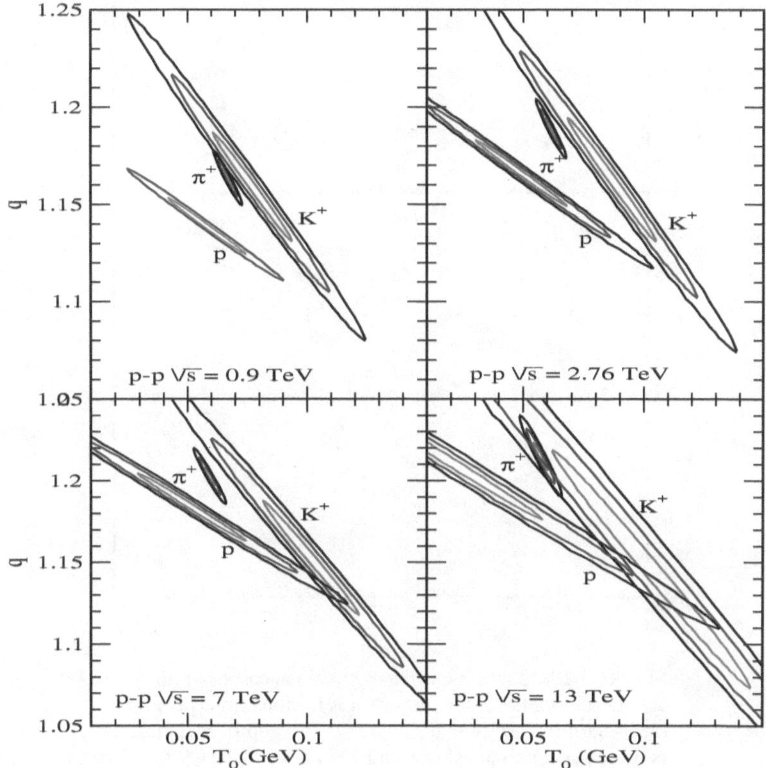

FIGURE 3.12: Contours in The $T_0 - q$ plane showing lines of 1 standard deviation from the minimum χ^2 in red. 2 standard deviations are shown in blue. Those for $3 - \sigma$ standard deviation from the minimum χ^2 are shown in black. The upper left panel is for data at $\sqrt{s} = 0.9$ TeV [69]; the upper right panel is for 2.76 TeV [28]. The lower left panel is for 7 TeV [29] while the lower right panel is for 13 TeV [70] with the CMS Collaboration.

From Figure 3.12, it can be seen that while the pions and kaons do overlap in a small region at $\sqrt{s} = 0.9, 2.76$ and 13 TeV, while protons and kaons also overlap at $\sqrt{s} = 7$ and 13 TeV.

Contours in the $T_0 - q$ plane showing 1 standard deviation from the minimum χ^2 are displayed in Figure 3.13 below:

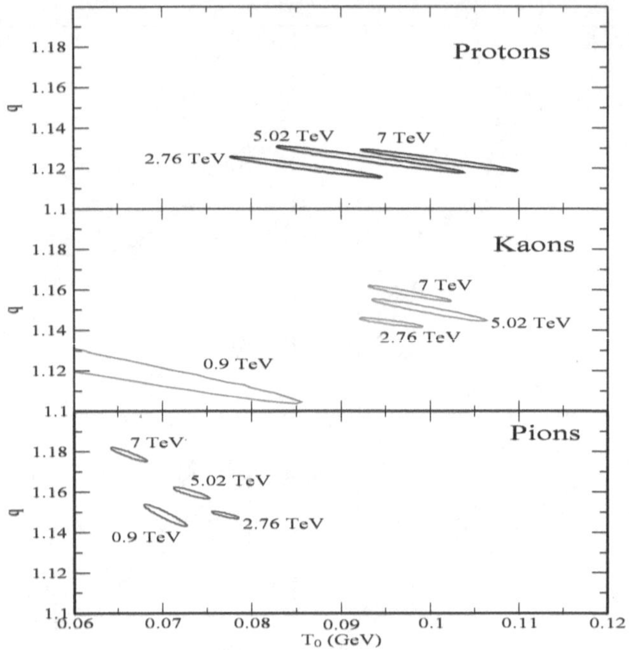

FIGURE 3.13: Contours at fixed χ^2 values corresponding to 1 standard deviations, for protons (top), kaons (middle) and pions (bottom). Notice that for the 0.9 TeV contours, pions and kaons respectively mean π^+ and K^+, and that the 0.9 TeV proton contour is simply out of range. The beam energies are displayed in the figure, and our results are obtained using data from the ALICE collaboration [65, 66, 67, 68].

Figure 3.13 shows that there is a clear beam energy dependence between 2.76 and 7 TeV given data from the ALICE collaboration. The results obtained at 0.9 TeV are out of line, presumably because of the limited range in p_T at this beam energy.

Discarding for the moment the results at 0.9 TeV, one can see a clear shift for the pions towards lower temperatures T_0 but higher values for q. For protons the opposite result is seen, namely a shift towards higher temperatures but the values for q are almost independent on the beam energy. For kaons the pattern is different again, namely an almost constant value for the temperature T_0 but a clear increase in the value of q.

Contours in the $T_0 - q$ plane showing 1 standard deviation from the minimum χ^2 are displayed in Figure 3.14 below:

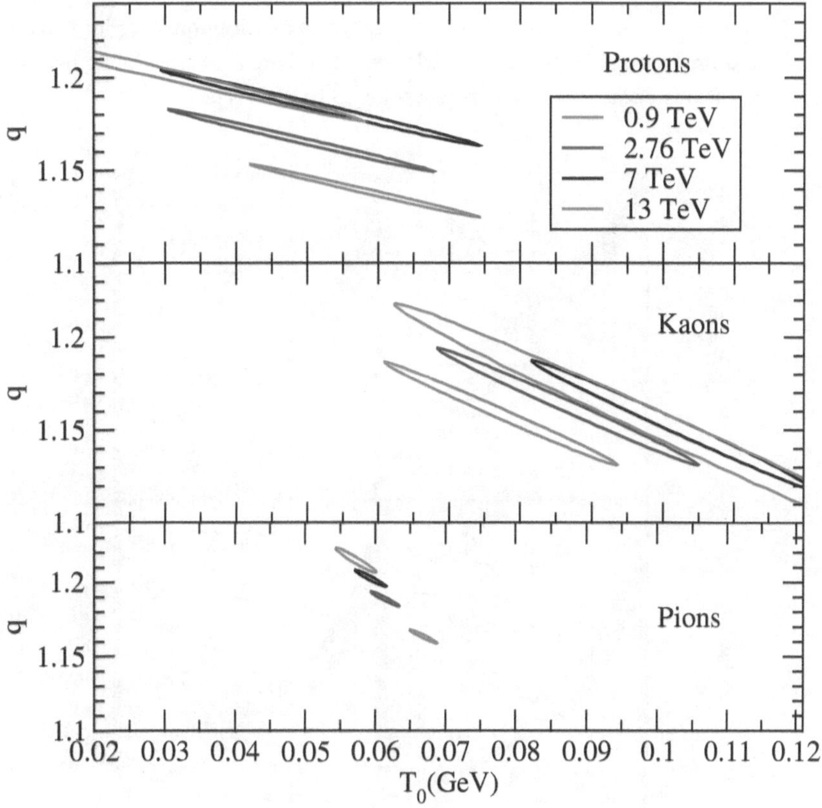

FIGURE 3.14: Contours at fixed χ^2 values corresponding to 1 standard deviations, for protons (top), kaons (middle) and pions (bottom). Notice that for the 0.9 TeV contours, pions and kaons respectively mean π^+ and K^+, and that the 0.9 TeV proton contour is simply out of range. The beam energies are displayed in the figure, and our results are obtained using data from the CMS collaboration [69, 70, 28, 29, 71]

The contours in Figure 3.14 do not show a clear beam energy dependence between. The results obtained at 13 TeV overlap for both protons and kaon at beam energies of 7 and 13 TeV. This does not have natural explanation apart from what we have identified as the weakness in fitting a very limited range in p_T.

In Figure 3.15 we show the transverse mass distributions for pions, kaons and protons at 2.76 TeV. It is clearly seen that there is no m_T scaling because the Tsallis parameters are different for each particle type.

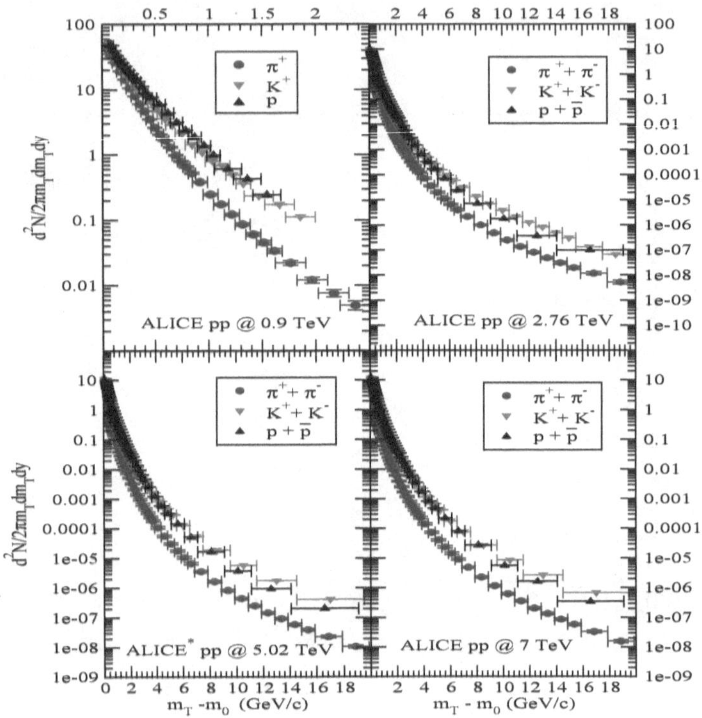

FIGURE 3.15:   Transverse mass distributions for π^+, K^+ and protons at 900 GeV (top left) [65]. Also shown are $\pi^+ + \pi^-$, $K^+ + K^-$ and $p + \bar{p}$ at 2.76 TeV (top right) [66] 5.02 TeV (bottom left) [67] and 7 TeV (bottom right) [68]. The data shown at 5.02 TeV are interpolated data.

We notice that our results clearly show that the fitted temperature T_0 is often comparable for kaons and for protons, which does not have a natural explanation in the radial flow scenario. In addition, the fitted q parameter is also often comparable for pions and kaons. As a consequence, different groups of nearly m_T scaling appear (see figure 3.15), groups whose characteristic allowing for differentiating them could well be the mass range of the given particles.

3.12 Summary

n this Chapter, we have determined the parameters appearing in the Tsallis distribution with the chemical potential set to zero as precisely as possible at beam energies of $\sqrt{s} = 0.9, 2.76, 5.02$ and 7 TeV for the ALICE data and $\sqrt{s} = 0.9, 2.76, 7$ and 13 TeV for the CMS data.

The conclusion we reach is that for the π's, K's and protons the parameters are different and no universality in the parameters exist. At the beam energy of $\sqrt{s} = 0.9$ TeV, the interval in transverse momentum is narrow and the uncertainty on the parameters is large. The suggestion, made a few years ago, that the parameters appearing in this distribution are the same for a wide range of identified hadrons [18, 19] at $\sqrt{s} = 900$ GeV in $p - p$ collisions is therefore not supported by the analysis presented here.

Thus, even though the Tsallis distribution provides a reasonable description of the transverse momenta distributions, it has to be concluded that the parameters are clearly different. As a consequence, one basic property of the Tsallis distribution, namely scaling in the transverse mass mT is not obeyed because the relevant parameters change for different hadrons.

Chapter 4

Tsallis fits with radial flow

This chapter combines Tsallis non-extensive statistics with radial flow. It has been argued that the inclusion of radial flow in the Tsallis distribution improves the description of data up to several values in p_T [72, 73, 74]. We seek to study the suitability of adding radial flow to the Tsallis distribution, in doing so, we would like to restrict our formulation to parameters that bear a physical meaning.

This work has been inspired by the recent successes in the literature including [37, 52, 72], however, questions about the specific form chosen for the flow remain unsolved in our opinion.

4.1   Model description

To combine the Tsallis prescription given in the previous chapter with flow, we follow the approach presented in [73]. Starting from eq. (2.15) , we replace the energy by $E \rightarrow p^\nu u_\nu$ where p^ν and u^ν are the particle 4-momentum and the collective flow 4-velocity respectively, and are given by [73]

$$
\begin{aligned}
u^\nu &= \gamma_r(\cosh\eta,\ v_r\cos\phi,\ v_r\sin\phi,\ \sinh\eta), \\
p^\nu &= (m_T\cosh y,\ p_T\cos\phi_P,\ p_T\sin\phi_P,\ m_T\sinh y),
\end{aligned}
\tag{4.1}
$$

here η is not pseudorapidity as previously defined, it is the space-time rapidity of the flow of the system: we did not change the symbol here to maintain the notation in [73]. The product $p^\nu u_\nu$ is given by

$$
p^\nu u_\nu = \gamma_r\left(m_T\cosh(y-\eta) - v_r p_T\cos(\phi_P - \phi)\right),
\tag{4.2}
$$

after simplification; here $\gamma_r = \frac{1}{\sqrt{1-v_r^2}}$ and v_r is the radial velocity. In the blast-wave model, one assumes instantaneous freeze-out which implies that τ_f is taken to be fixed [72, 73] and the freeze-out hyper surface is parametrised as

$$d^3\Sigma_\nu = \tau_f(\cosh\eta,\ 0,\ 0,\ \sinh\eta)d\eta\ d^2r. \tag{4.3}$$

The proper freeze-out time τ_f could depend on the distance from the centre of the expanding fireball. The particle distribution is now given by the Cooper-Frye [75, 76] formula

$$E\frac{d^3N}{d^3p}\bigg|_{y=0} = \frac{g}{(2\pi)^3}\int d^3\Sigma_\nu\ p^\nu f(x), \tag{4.4}$$

and

$$p^\nu d^3\Sigma_\nu = m_T\tau_f\cosh(y-\eta)d\eta\ d^2r, \tag{4.5}$$

it follows that

$$E\frac{d^3N}{d^3p}\bigg|_{y=0} = \frac{g}{(2\pi)^3}\tau_f m_T\int_{-\eta_{max}}^{+\eta_{max}} d\eta\ \cosh(y-\eta)$$
$$\int_0^R rdr\int_0^{2\pi} d\phi\left[1+(q-1)\frac{p^\nu u_\nu - \mu}{T}\right]^{-\frac{q}{q-1}}. \tag{4.6}$$

In the Bjorken model where there is a plateau in the rapidity, one introduces a Dirac delta distribution function $\delta(y-\eta)$ such that the integral over rapidity dependence disappears: thus, for Bjorken scaling which is independent of rapidity, one has [77]

$$E\frac{d^3N}{d^3p}\bigg|_{y=0} = \frac{g}{(2\pi)^3}\tau_f m_T\int_0^R rdr$$
$$\int_0^{2\pi} d\phi\left[1+(q-1)\frac{p^\nu u_\nu - \mu}{T}\right]^{-\frac{q}{q-1}}. \tag{4.7}$$

Now substituting for $p^\nu u_\nu$ and at zero chemical potential, we have

$$E\frac{d^3N}{d^3p}\bigg|_{y=0} = \frac{g}{(2\pi)^3}\tau_f m_T\int_0^R rdr$$
$$\int_0^{2\pi} d\phi\left[1+(q-1)\frac{\gamma_r\left(m_T - v_r p_T\cos(\phi_P - \phi)\right)}{T}\right]^{-\frac{q}{q-1}} \tag{4.8}$$

Now for the integral over ϕ, we let

$$I \equiv \int_0^{2\pi} d\phi\left[1+\frac{(q-1)}{T}\gamma_r m_T - \frac{(q-1)}{T}\gamma_r v_r p_T\cos(\phi_P - \phi)\right]^{-\frac{q}{q-1}}. \tag{4.9}$$

Here, we can shift $(\phi_P - \phi)$ to simply ϕ as $d\phi \to d(\phi_P - \phi)$. Now we can use the Legendre polynomial $P_n(z)$ which is given by [78]

$$
\begin{aligned}
P_n(z) &= \frac{1}{\pi} \int_0^\pi \left(z - \sqrt{z^2 - 1} \cos t \right)^n dt \\
&= \frac{z^n}{\pi} \int_0^\pi \left(1 - \frac{\sqrt{z^2 - 1}}{z} \cos t \right)^n dt,
\end{aligned}
\tag{4.10}
$$

where n is not an integer, and we bring the factor independent of ϕ outside the integral such that

$$
I = \left[1 + \frac{(q - 1)}{T} \gamma_r m_T \right]^{-\frac{q}{q-1}} \int_0^{2\pi} d\phi \left[1 - \frac{\frac{q-1}{T} \gamma_r v_r p_T}{\left[1 + \frac{q-1}{T} \gamma_r m_T \right]} \cos \phi \right]^{-\frac{q}{q-1}} .
\tag{4.11}
$$

We identify the coefficient of $\cos(\phi)$ with

$$
\frac{\sqrt{z^2 - 1}}{z} = \frac{\frac{q-1}{T} \gamma_r v_r p_T}{1 + \frac{q-1}{T} \gamma_r m_T} \equiv b.
\tag{4.12}
$$

This, the integral becomes

$$
I = \left[1 + \frac{(q - 1)}{T} \gamma_r m_T \right]^{-\frac{q}{q-1}} \frac{2\pi}{z^n} P_n(z).
\tag{4.13}
$$

Now, following the prescription in [73], we call $z \equiv a$, we have

$$
E \frac{d^3 N}{d^3 p} \bigg|_{y=0} = \frac{g}{(2\pi)^3} \tau_f m_T \int_0^R r\, dr \left[1 + \frac{(q - 1)}{T} \gamma_r m_T \right]^{-\frac{q}{q-1}} 2\pi\, a^{\frac{q}{q-1}} P_{-\frac{1}{q-1}}(a).
\tag{4.14}
$$

For negative indices of the Legendre polynomial $P_n = P_{-n-1}$ or $P_{-n} = P_{n-1}$, which implies that $\frac{q}{q-1} - 1 = \frac{q-q+1}{q-1} = \frac{1}{q-1}$. substituting back gives

$$
E \frac{d^3 N}{d^3 p} \bigg|_{y=0} = \frac{g}{(2\pi)^2} \tau_f m_T \int_0^R r\, dr \left[1 + \frac{(q - 1)}{T} \gamma_r m_T \right]^{-\frac{q}{q-1}} a^{\frac{q}{q-1}} P_{\frac{1}{q-1}}(a).
\tag{4.15}
$$

Taking $v_r(r) = v_s(\frac{r}{R})^2$ [73], and defining $a \equiv \frac{1}{\sqrt{1-b^2}}$ with

$$
b = \frac{\frac{(q-1)}{T} \gamma_r v_r p_T}{1 + \frac{(q-1)}{T} \gamma_r m_T},
\tag{4.16}
$$

we can re-write eq. (4.15) as

$$E\frac{d^3N}{d^3p}\bigg|_{y=0} = \frac{g}{(2\pi)^2}\tau_f m_T \int_0^R rdr \left[\frac{a}{1+\frac{(q-1)}{T}\gamma_r m_T}\right]^{\frac{q}{q-1}} P_{\frac{1}{q-1}}(a). \qquad (4.17)$$

The above equation can be implemented in ROOT using the available numerical schemes.

4.2   Fits with radial flow

The Tsallis parameters were extracted by fitting the experimental results published by the ALICE collaboration [65, 66, 67, 68] to eq. (4.17) and the results are presented below.

TABLE 4.1: The extracted values of v_r, q, T_0 and χ^2/NDF for $p-p$ collisions at $\sqrt{s} = 900$ GeV using data from the ALICE collaboration [65].

Particle	v_r	q	T_0 (MeV)	χ^2/NDF
π^+	0.6808 ± 0.2177	1.129 ± 0.017	54.5 ± 13.5	$19.1769/29$
π^-	0.6650 ± 0.2228	1.125 ± 0.018	56.9 ± 13.0	$14.3605/29$
K^+	0.3486 ± 0.1910	1.182 ± 0.009	39.8 ± 11.8	$22.9792/23$
K^-	0.3773 ± 0.1108	1.166 ± 0.011	47.2 ± 9.0	$10.8172/23$
p	0.0029 ± 0.9791	1.147 ± 0.005	32.7 ± 4.2	$36.2781/20$
$\bar{p}$	0.0220 ± 0.6346	1.132 ± 0.0124	51.9 ± 17.2	$36.262/20$

In Table 4.1 , the resulting central values of T_0 are lower than the ones presented in Table 3.1 for all the particles while χ^2/NDF values are much higher. The extracted flow value e.g. for protons, $v_r = 0.0029 \pm 0.9791$ is unrealistic, the same applies to anti-protons. Such a result makes it difficult to draw conclusions or infer meaning from these fit results.

The transverse momentum distributions at beam energy of $\sqrt{s} = 0.9$ TeV are shown in Figure 4.1 where it can be seen that for all particles, the fits describe the data very well over the entire p_T range.

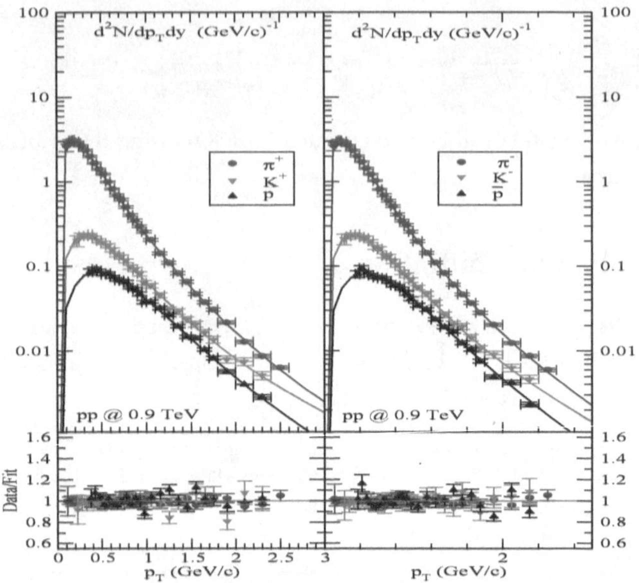

FIGURE 4.1: Fits to the transverse p_T distributions, using the Tsallis distribution eq. (4.17). The ALICE data at $\sqrt{s} = 0.9$ TeV are as follows: pions are in blue, kaons in red and protons in black. The solid curves represent the fit to data. The Data/Fit ratios are presented below the corresponding p_T distributions.

TABLE 4.2: The extracted values of v_r, q, T_0 and χ^2/NDF for $p - p$ collisions at $\sqrt{s} = 2.76$ TeV using data from the ALICE collaboration [66].

Particle	v_r	q	T_0 (MeV)	χ^2/NDF
$\pi^+ + \pi^-$	0.8876 ± 0.0495	1.141 ± 0.002	30.3 ± 9.0	$17.647/60$
$K^+ + K^-$	0.0062 ± 0.5214	1.144 ± 0.002	95.6 ± 4.5	$14.4217/55$
$p + \bar{p}$	0.0045 ± 0.6912	1.121 ± 0.005	85.8 ± 8.2	$34.736/46$

In Table 4.2, the resulting central values of T_0 is much lower for pions than the ones presented in Table 3.2 and is the same for kaons and protons. The results for radial flow are unrealistic for protons and kaons which are compatible with zero.

The trend in fit results presented Tables 4.3 and 4.4 are the same as the ones reported in the previous two tables in that the resulting values for flow

TABLE 4.3: The extracted values of v_r, q, T_0 and χ^2/NDF for $p - p$ collisions at $\sqrt{s} = 5.02$ TeV using data from the ALICE collaboration [67].

Particle	v_r	q	T_0 (MeV)	χ^2/NDF
$\pi^+ + \pi^-$	0.90995 ± 0.03361	1.148 ± 0.003	25.8 ± 7.8	$14.6188/55$
$K^+ + K^-$	0.4433 ± 0.4008	1.150 ± 0.005	79.0 ± 4.6	$12.2452/48$
$p + \bar{p}$	0.00116 ± 0.5291	1.126 ± 0.005	91.2 ± 8.6	$20.7735/46$

TABLE 4.4: The extracted values of v_r, q, T_0 and χ^2/NDF for $p - p$ collisions at $\sqrt{s} = 7.0$ TeV using data from the ALICE collaboration [68].

Particle	v_r	q	T_0 (MeV)	χ^2/NDF
$\pi^+ + \pi^-$	0.9179 ± 0.02751	1.151 ± 0.003	24.1 ± 6.9	$20.2327/55$
$K^+ + K^-$	0.4108 ± 0.3502	1.154 ± 0.005	83.5 ± 34.7	$27.6692/48$
$p + \bar{p}$	0.0033 ± 0.8946	1.129 ± 0.005	93.7 ± 9.5	$29.9594/46$

are unrealistic with error bars much larger than or almost equal to the central values.

4.3 Analysis of results

Following the work of [37], who analysed only pions and quarkonia; if one focuses only on the flow results for pions presented in this chapter, one will conclude that the formulation presented here is successful in extracting the flow from the transverse momentum spectra.

The results presented in this chapter show that adding flow to the Tsallis distribution reproduces the transverse momentum spectra. However, the resulting flow values are unrealistic for protons and kaons. That is, flow cannot be determined and consistent with zero (no flow) for protons and in Table 4.3 and Table 4.4 flow almost equal the speed of light. These values are higher than those for a massless gase. Usually, the maximum flow in a massless gas equals $1/\sqrt{3}$.

Chapter 5

Tsallis fits with chemical potential

In this chapter we consider the possibility of including the chemical potential parameter in the Tsallis fits to the transverse momentum spectra. Previously, it was first noted by [47] that the variables T, V, q, μ in the Tsallis distribution function eq. (2.15) have a redundancy for $\mu \neq 0$ and recently [79] considered the mass of a particle in place of chemical potential. This necessitates work on determining the chemical potential from the transverse momentum spectra.

5.1   Model choice

In order to extract the chemical potential from the transverse momentum spectra, we make use of eq. (2.16) reprinted below

$$\left.\frac{d^2N}{dp_T dy}\right|_{y=0} = gV\frac{p_T m_T}{(2\pi)^2}\left[1+(q-1)\frac{m_T-\mu}{T}\right]^{-\frac{q}{q-1}}.$$

We then extract the parameters (T, q, μ and V) which we then utilise to calculate the corresponding value of T_0 from eq. (2.17). Another approach which one can utilise in determining the chemical potential was also proposed by [61], where the observation was made that the radius R_0 given in Table 5.3 is larger than the one obtained from a femtoscopy analysis [80] by a factor κ estimated to be about 3.5, i.e.,

$$R_{\text{femto}} \approx \frac{1}{\kappa}R_0. \tag{5.1}$$

Hence in [61] the suggestion is made to identify the corresponding volume V_{femto} with the volume V appearing in Equation (2.15). Hence

$$V_0 \approx V \cdot \kappa^3. \tag{5.2}$$

Combining e.q (5.2) together with e.qs (2.17) and (2.18) leads to a chemical potential given by

$$\mu = \frac{T_{\mu=0}}{q-1}\left(\kappa^{3(q-1)/q} - 1\right). \tag{5.3}$$

We shall utilise all these proposals to extract the chemical potential from the transverse momentum spectra.

5.2   Comparison of fit results

In the first case, we set the chemical potential as the mass of the respective particle and compare our results to [79] for $p - p$ collisions at 900 GeV with the CMS collaboration, secondly we set the chemical potential as a free parameter to fit the data and analysis of the fit results and lastly, we calculate the chemical potential directly from eq. (5.3).

A comparison of the values of temperature for different hadron species for $p - p$ collisions at $\sqrt{s} = 900$ with CMS [28] collaboration is shown in Figure 5.1. The results show that the temperature increases with particle mass: this is different from the results previously published by [18] which shows that the temperature is the same for all particles within errors.

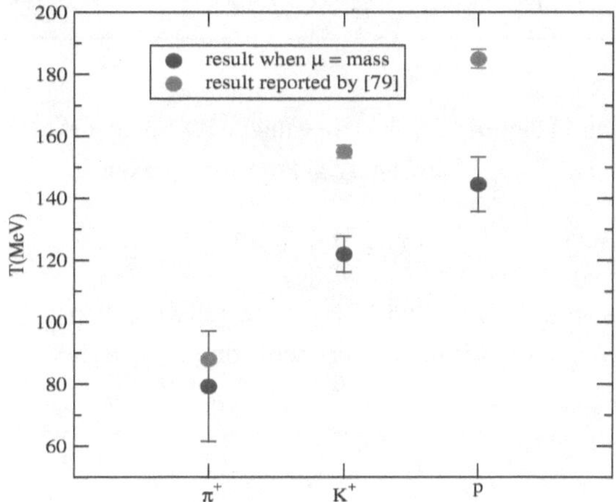

FIGURE 5.1:   A comparison of the values of temperature for different hadron species for $p - p$ collisions $\sqrt{s} = 900$ with CMS [28] collaboration. The result reported by [79] is represented by red circles while the result obtained by setting the chemical potential equal to the respective particle mass are represented by blue circles.

In the comparison above, there is a systematic difference in the temperature to the values reported by [79] for $p - p$ collisions at $\sqrt{s} = 900$ with CMS [28] collaboration, however, the trend in the result is the same. At this stage, it is difficult to make a deduction regarding the chemical potential from this result and we note also that using the mass in place of the chemical potential destroys thermodynamic consistency.

In Table 5.1, we present the fit results for non-zero chemical potential for $p - p$ collisions at $\sqrt{s} = 900\,\text{GeV}$ with the ALICE experiment. We notice that the chemical potential is the same within errors for each particle antiparticle pair: a scenario expected at LHC energies, given that our analysis is done at kinetic freeze-out.

TABLE 5.1: The extracted values of T, q, R μ and χ^2/NDF parameters, using the data published in [65] for $p - p$ collisions at $\sqrt{s} = 900\,\text{GeV}$ with the ALICE experiment.

Particle	T (MeV)	q	R (fm)	μ (MeV)	χ^2/NDF
π^+	97 ± 3	1.148 ± 0.0051	3.873 ± 0.215	181 ± 14	$22.73/29$
K^+	81 ± 4	1.175 ± 0.0108	3.768 ± 0.309	139 ± 27	$13.02/23$
p^+	76 ± 4	1.161 ± 0.0090	3.697 ± 0.278	367 ± 30	$14.52/20$
π^-				180 ± 1	$18.17/32$
K^-				139 ± 3	$12.24/26$
$\bar{p}$				361 ± 23	$19.19/23$

The results in Figure 5.2 show that values of T_0 calculated using the values of T, q and μ in Table 5.1 and eq. (2.17) reprinted below

$$T_0 = T \left[1 - (q - 1)\frac{\mu}{T}\right],$$

gives consistent values of T_0 and this confirms the assertion by [47] that there exist a redundancy in the parameters appearing in eq. (2.16).

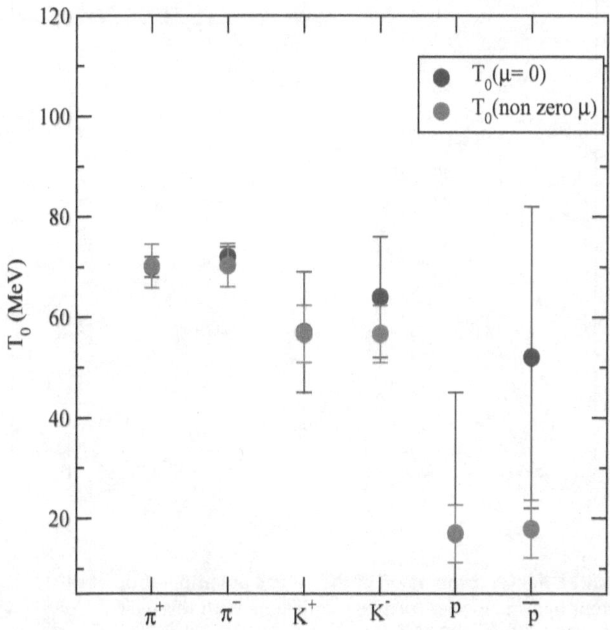

FIGURE 5.2: A comparison of the values of temperature of different hadron species for $p - p$ collisions $\sqrt{s} = 900$ GeV with the ALICE [65] collaboration. The temperature values calculated by eq. (2.17) are represented by red circles while the results presented in Table 5.1 are represented by blue circles.

The results in Figure 5.3 show that values calculated by eq. (2.17) are in agreement with those extracted when the chemical potential is set to zero, and this confirms an assertion by [47] that there exist a redundancy in the parameters appearing in eq. (2.16).

For the result at $\sqrt{s} = 2.76$ TeV, the fit routine gave huge errors on the extracted chemical potential, and this resulted in huge errors on the calculated temperature, however, the central values are in agreement.

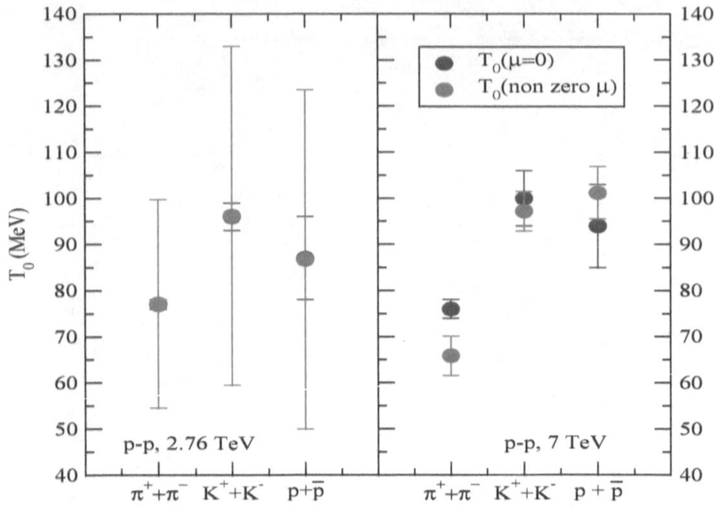

FIGURE 5.3: A comparison of the values of temperature of different hadron species for $p - p$ collisions with the ALICE Collaboration at $\sqrt{s} = 2.76$ TeV [66] on the left panel and at $\sqrt{s} = 7$ TeV with [68] on the right. The temperature values calculated by eq. (2.17) are represented by red circles while the results presented in Table 3.2 and Table 3.4 are represented by blue circles for the respective energies.

In Table 5.2 we present the extracted values of T, q, R μ at four different energies with the CMS collaboration. For the antiparticles, we fix the parameters T, q and R equal to those for the corresponding particle type, since we wanted to compare the values of μ for each pair. The results presented here show that indeed for each pair, the chemical potential values are the same within errors.

TABLE 5.2: The extracted values of T, q, R, μ and χ^2/NDF parameters, using the data published in [69, 70, 71] for $p-p$ collisions with the CMS experiment.

$\sqrt{s}$ (GeV)	Particle	T (MeV)	q	R (fm)	μ (MeV)	χ^2/NDF
900 [69]	π^+	95 ± 1	1.164 ± 0.004	4.344 ± 0.102	173 ± 6	4.044/18
	K^+	93 ± 2	1.158 ± 0.008	3.852 ± 0.117	101 ± 12	2.123/13
	p^+	92 ± 2	1.139 ± 0.003	4.03 ± 0.110	246 ± 12	9.596/23
	π^-				171 ± 1	11.01/21
	K^-				98 ± 1	1.974/16
	$\bar{p}$				233 ± 1	30.18/26
2760 [71]	π^+	96 ± 1	1.189 ± 0.005	4.439 ± 0.104	185 ± 6	5.711/18
	K^+	100 ± 3	1.162 ± 0.008	3.971 ± 0.121	80 ± 13	2.447/13
	p^+	93 ± 2	1.166 ± 0.004	3.802 ± 0.101	262 ± 13	27.43/23
	π^-				183 ± 1	8.534/21
	K^-				77 ± 1	7.653/16
	$\bar{p}$				255 ± 1	36/26
7000 [71]	π^+	100 ± 2	1.203 ± 0.005	4.481 ± 0.108	199 ± 6	14.29/18
	K^+	111 ± 3	1.152 ± 0.009	3.985 ± 0.122	56 ± 14	2.074/13
	p^+	96 ± 2	1.184 ± 0.004	3.808 ± 0.101	238 ± 14	12.22/23
	π^-				197 ± 1	11.69/21
	K^-				54 ± 1	5.279/16
	$\bar{p}$				236 ± 1	15.7/26
13000 [70]	π^+	103 ± 2	1.215 ± 0.008	4.129 ± 0.152	211 ± 10	3.546/18
	K^+	110 ± 5	1.142 ± 0.0150	4.125 ± 0.24	29 ± 3	1.828/13
	p^+	85 ± 3	1.213 ± 0.008	3.617 ± 0.143	313 ± 20	8.892/22
	π^-				209 ± 1	25.36/21
	K^-				25 ± 3	1.478/16
	$\bar{p}$				306 ± 2	9.274/25

The results in Figure 5.4 show that values calculated by eq. (2.17) are in agreement with those extracted when the chemical potential is set to zero, this further confirms an assertion by [47] that there exist a redundancy in the parameters appearing in eq. (2.16).

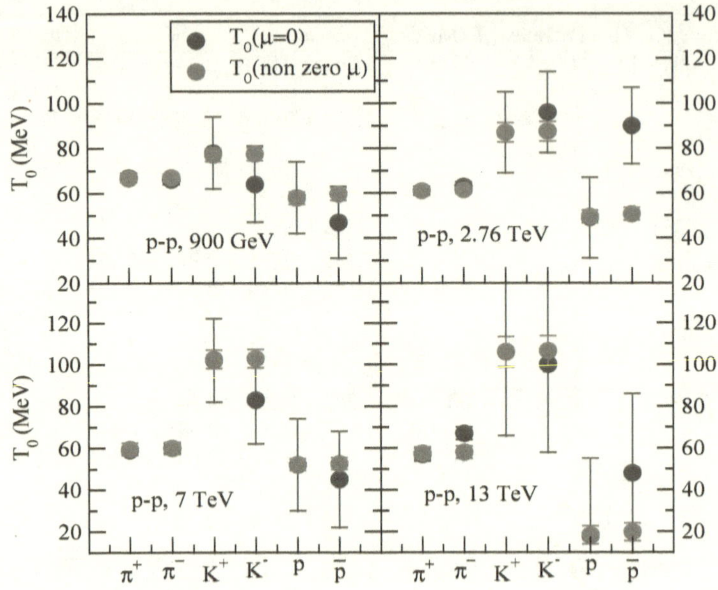

FIGURE 5.4: A comparison of the values of temperature of different hadron species for $p-p$ collisions at $\sqrt{s} = 0.9, 2.76, 7$ and 13 TeV [69, 70, 71] with the CMS Collaboration as indicated in the different panels above. The temperature values calculated by eq. (2.17) are represented by red circles while the results presented in Tables 3.5, 3.6, 3.7 and Table 3.8 are represented by blue circles for the respective energies.

The results in Figure 5.5 show that values calculated by eq. (5.3) are not in agreement with those extracted when the chemical potential is set to zero. In coming up with eq. (5.3), the authors in [61] made an assumption that $V_{\mu=0} = V_\mu \cdot \kappa^3$, which together with eq. (2.17) and eq. (2.17) lead to the formulation of eq. (5.3); hence, we fail to confirm this assumption.

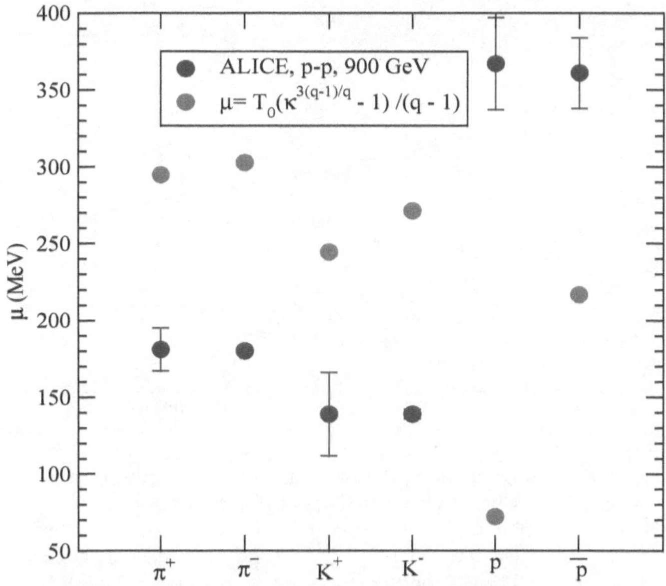

FIGURE 5.5: Chemical potential as a function of particle type. The blue circles are values of chemical potential extracted using data in $p - p$ collisions at 900 GeV with the ALICE collaboration, see Tables 5.1. The red circles are the values calculated by eq. (5.3).

TABLE 5.3: The extracted values of T_0, q, R_0 and χ^2/NDF parameters, using the data published in [81] for $p - p$ collisions with the NA 61 Collaboration.

$\sqrt{s}$ (GeV)	Particle	T_0 (MeV)	q	R_0 (fm)	χ^2/NDF
6.3	π^-	98 ± 6	1.042 ± 0.015	2.55 ± 0.14	$4.454/15$
7.7	π^-	95 ± 3	1.057 ± 0.008	2.72 ± 0.09	$4.561/15$
8.8	π^-	96 ± 2	1.055 ± 0.006	2.76 ± 0.06	$8.423/15$
12.3	π^-	95 ± 2	1.064 ± 0.006	2.90 ± 0.06	$6.775/15$
17.3	π^-	93 ± 3	1.069 ± 0.006	3.07 ± 0.08	$2.176/15$

The results in Figure 5.6 show that the extracted T_0 values are in agreement with those given by the predicted by $0.1014\sqrt{s}^{-0.03262}$ [61]. However,

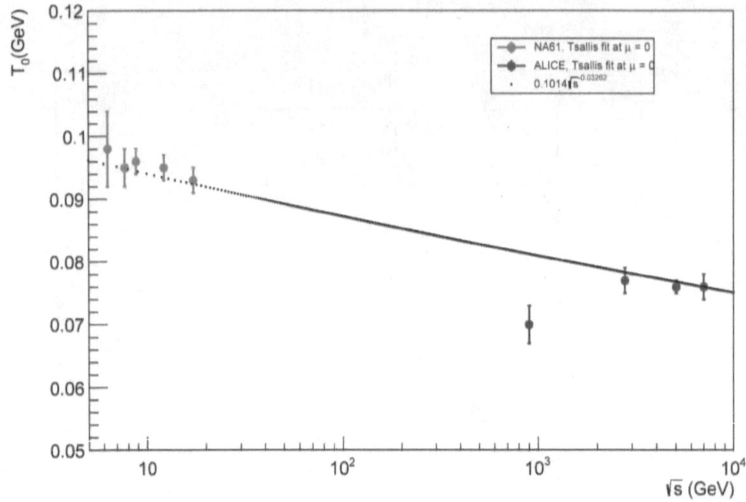

FIGURE 5.6: The energy dependence of the temperature parameter T_0. The blue circles are values of T extracted at zero chemical potential in $p - p$ collisions with the ALICE collaboration for pions. (see Tables 3.1, 3.2, 3.3 and and 3.4). These are fitted by eq. (5.3)[61] and the blue circles are the T_0 values in Table 5.3.

at 900 GeV, we obtain a much smaller value in $p - p$ collisions at 900 GeV.

5.3 Summary

A comparison of T and T_0 values for all the energies considered, for both the ALICE and CMS Collaborations are in agreement. This result confirms that the variables T, V, q, and μ in the Tsallis distribution function eq. (2.15) have a redundancy for $\mu \neq 0$ [47].

The fit results presented here clearly show that the chemical potential is the same for each particle and its antiparticle at each of the energies considered, a scenario expected at LHC energies.

Chapter 6

Tsallis fits to Pb-Pb collisions

6.1   Introduction

This chapter presents the application of Tsallis distribution in the analysis of
transverse momentum distributions in Pb-Pb collisions. The data we make
use of are published by the ALICE collaborations [66, 82] and by the NA 49
collaboration [83, 84]. Part of results presented in this chapter were presented
in [1]. It should be noted that in Figures 6.6 to 6.9, the results of an analysis
of Xe-Xe collisions are included: but not in [1]; they were only mentioned in
this article.

Firstly, we present the fits to transverse mass spectra for identified hadrons,
this is followed by the fits to transverse momentum spectra for charged hadrons.
Following this, we utilise the single particle distribution to determine the
temperature T_0 and the Tsallis parameter q.

These extracted parameters are then utilised to calculate the corresponding thermodynamic quantities namely, the energy density, the pressure and
entropy density. The values obtained for these thermodynamic quantities
are then discussed and compared to values obtained at different stages of the
collision and to other closely related energy densities.

6.2   Transverse mass spectra for identified hadrons

The transverse mass distributions of identified hadrons produced in Pb-Pb
collisions with the NA 49 collaboration at CERN are fitted using a Tsallis
distributions eq. (2.20). In Table 6.1 , we present the results for temperature
T_0 and the Tsallis parameter q at different center of mass energies.

TABLE 6.1: The extracted values of T_0, q and χ^2/NDF parameters, using the for the Pb-Pb collisions at 20, 30 GeV [83], 40, 80 and 158 GeV [84] with the NA 49 Collaboration.

$\sqrt{s}$ (GeV)	Particle	T_0 (MeV)	q	χ^2/NDF
	π^+	50 ± 1	1.215 ± 0.003	$204.2/13$
6.3	π^-	39 ± 1	1.268 ± 0.004	$124/13$
	K^+	75 ± 2	1.22 ± 0.009	$150.7/7$
	K^-	63 ± 2	1.253 ± 0.016	$23.6/7$
	π^+	49 ± 1	1.242 ± 0.004	$278.5/13$
7.7	π^-	40 ± 1	1.286 ± 0.004	$212.3/13$
	K^+	74 ± 1	1.26 ± 0.009	$223/7$
	K^-	46 ± 1	1.379 ± 0.011	$413.1/7$
	π^-	49 ± 1	1.257 ± 0.008	$68.69/11$
8.8	K^+	66 ± 1	1.28 ± 0.005	$573.3/7$
	K^-	77 ± 2	1.207 ± 0.012	$88.41/7$
	π^-	44 ± 1	1.283 ± 0.008	$45.02/11$
12.3	K^+	56 ± 1	1.336 ± 0.010	$215.4./11$
	K^-	52 ± 1	1.374 ± 0.008	$324.1/15$
17.3	π^-	43 ± 1	1.292 ± 0.009	$43/11$

For all the different center of mass energies considered here, the extracted parameter values stay largely the same, albeit small differences, this makes it difficult to tease out the variation of these parameters with increasing center of mass energy.

For the π^- particle, we have the data at all different energies and analyse fits for π^- particle with increasing energy: the results are presented in Figure 6.1 which indicates that the fits reproduces the data very well.

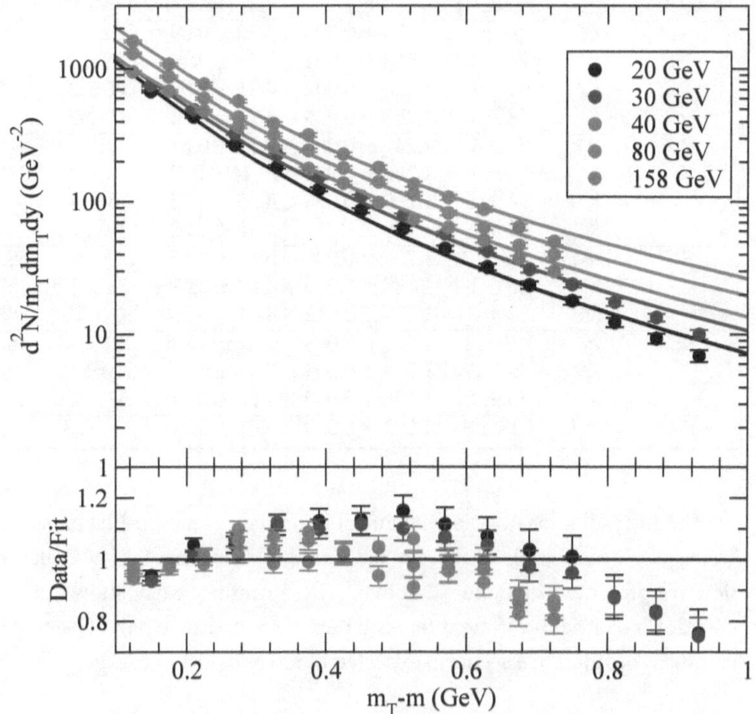

FIGURE 6.1: Transverse mass distributions for the π^- particle measured by the NA49 collaboration in Pb-Pb collisions at at 20, 30 GeV [83], 40, 80 and 158 GeV [84] as a function of the center of mass energy. The lower part of the figure shows the ratio of the data divided by the fit value.

TABLE 6.2: The extracted values of T, q, R, μ and χ^2/NDF parameters, using the for the Pb-Pb collisions at $20, 30$ GeV [83], $40, 80$ and 158 GeV [84] with the NA 49 Collaboration.

$\sqrt{s}$ (GeV)	type	T (MeV)	q	R (fm)	μ (MeV)	χ^2/NDF
6.3	π^+	106 ± 1	1.215 ± 0.003	4.20 ± 0.09	262 ± 5	$204.2/12$
	π^-	76 ± 1	1.268 ± 0.004	7.46 ± 0.19	138 ± 5	$124/12$
	K^+	91 ± 1	1.220 ± 0.009	4.26 ± 0.02	71 ± 2	$150.7/6$
	K^-	68 ± 1	1.253 ± 0.035	4.23 ± 0.04	16 ± 2	$23.6/6$
7.7	π^+	73 ± 1	1.242 ± 0.004	9.32 ± 0.28	99 ± 5	$278.5/12$
	π^-	50 ± 1	1.286 ± 0.004	15.44 ± 0.44	35 ± 3	$212.3/12$
	K^+	94 ± 1	1.260 ± 0.008	4.38 ± 0.02	78 ± 1	$223/6$
	K^-	61 ± 3	1.379 ± 0.011	4.43 ± 0.02	39 ± 7	$413.1/6$
8.8	π^-	105 ± 2	1.257 ± 0.008	5.52 ± 0.15	218 ± 7	$68.69/10$
	K^+	86 ± 1	1.280 ± 0.005	4.63 ± 0.01	74 ± 1	$573.3/6$
	K^-	83 ± 3	1.207 ± 0.012	4.04 ± 0.26	30 ± 13	$88.41/6$
12.3	π^-	101 ± 1	1.283 ± 0.008	6.42 ± 0.018	201 ± 1	$45.02/10$
	K^+	97 ± 3	1.336 ± 0.009	3.97 ± 0.17	120 ± 9	$215.4./10$
	K^-	73 ± 1	1.374 ± 0.008	4.30 ± 0.01	57 ± 1	$324.1/14$
17.3	π^-	108 ± 3	1.292 ± 0.009	6.41 ± 0.21	220 ± 8	$43/10$

The fit results for non-zero chemical potential for the Pb-Pb collisions at $\sqrt{s} = 20, 30$ GeV [83], $40, 80$ and 158 GeV [84] with the NA 49 Collaboration are presented in Table 6.2 above. These results clearly shows that the chemical potential is different for each particle antiparticle pair at each of the energies considered, a scenario expected at lower collision energy.

In Figure 6.2, we present a comparison of the temperature values extracted with the chemical potential set to zero, see Table 6.1, to the temperature values calculated by eq. (2.17). For all the energies considered, the calculated values are in agreement with the values extracted when the chemical is set to zero, and this result further confirms that the variables T, V, q, μ in the Tsallis distribution function eq. (2.15) have a redundancy for $\mu \neq 0$ [47].

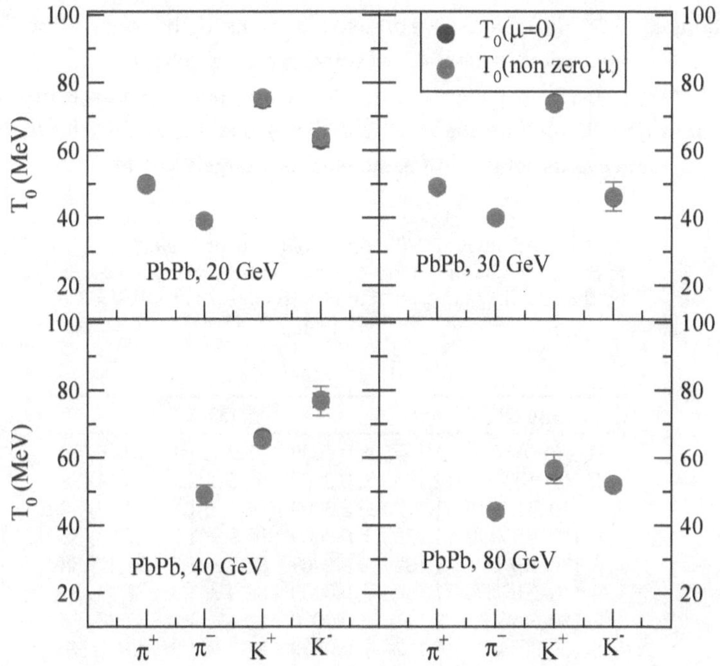

FIGURE 6.2: A comparison of the values of temperature of different hadron species for Pb-Pb collisions at $\sqrt{s} = 20, 30$ GeV [83], 40, 80 and 158 GeV [84] with the NA 49 Collaboration as indicated in the different panels above. The temperature values calculated by eq. (2.17) are represented by red circles while the results presented in Table 6.1 are represented by blue circles for the respective energies.

6.3 Transverse momentum spectra for charged hadrons

The transverse momentum distributions of charged hadrons produced in Pb-Pb collisions at LHC energies are fitted using a sum of three Tsallis distributions eq. (2.23). In Table 6.3, we present the results for the temperature T_0 and the Tsallis parameter q for all nine centrality classes obtained from fitting the Pb-Pb data at a beam energy of 2.76 TeV. Here we notice that the temperature varies from 96 MeV for the most central events and to 78 MeV for the most peripheral events, while the q parameter stays largely at 1.14.

TABLE 6.3: Values of q, T and χ^2/NDF obtained using eq. (2.24) to fit charged hadron transverse momentum spectra measured by the ALICE collaboration in Pb-Pb collisions at $\sqrt{s_{NN}} = 2.76$ TeV [85].

Centrality Class	q	T_0 (MeV)	χ^2/NDF
1 (0-5)%	1.1355 ± 0.0009	95.9 ± 1.4	156.5/58
2 (5-10)%	1.1363 ± 0.0009	95.5 ± 1.3	150.4/58
3 (10-20)%	1.1376 ± 0.0009	94.5 ± 1.3	137.9/58
4 (20-30)%	1.1387 ± 0.0009	92.9 ± 1.3	117.3/58
5 (30-40)%	1.1389 ± 0.0009	91.2 ± 1.3	91.47/58
6 (40-50)%	1.1403 ± 0.0009	88.0 ± 1.3	71.39/58
7 (50-60)%	1.1416 ± 0.0010	84.6 ± 1.3	52.88/58
8 (60-70)%	1.1424 ± 0.0010	81.0 ± 1.3	29.8/58
9 (70-80)%	1.1428 ± 0.0012	78.0 ± 1.3	23.16/58

Following the same procedure, we present in Table 6.3 the results for the temperature T_0 and the Tsallis parameter q for all nine centrality classes obtained from fitting the Pb-Pb data at a beam energy of 2.76 TeV. Here we notice that the temperature varies from 96 MeV for the most central events and to 78 MeV for the most peripheral events, while the q parameter mostly stay around 1.14.

The resulting fits to the experimental data obtained in Pb-Pb collisions at $\sqrt{s_{NN}} = 2.76$ TeV are shown in Figure 6.3 where we follow the centrality classification introduced in [85]. As can be seen in Figure 6.3 the fits are very good for peripheral events and at low p_T, gradually worsening for the more central events where the fits at first overshoot the data above p_T values of about 3 GeV then rejoin the data and at larger values of p_T above about 20 GeV are below the data.

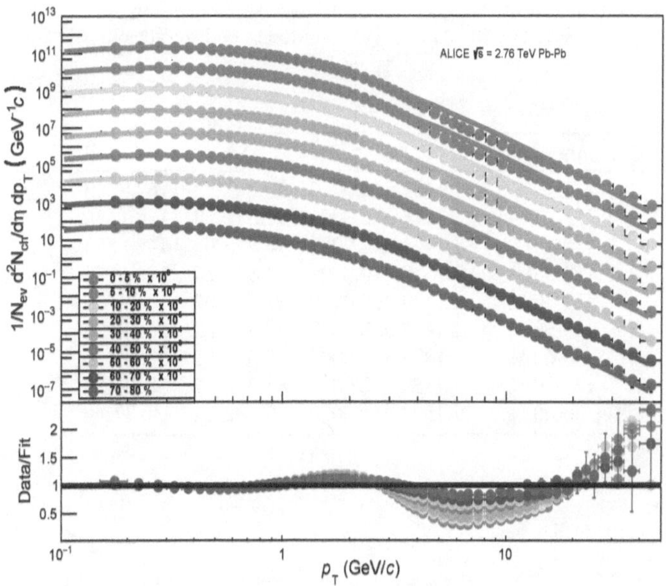

FIGURE 6.3: Transverse momentum distributions measured by the ALICE collaboration in Pb-Pb collisions at $\sqrt{s_{NN}}$ = 2.76 TeV [85] as a function of centrality. The solid lines represent fits made using the Tsallis distribution eq. 2.23. The lower part of the figure shows the ratio of the data divided by the fit value.

Following the same procedure as before, we present in Table 6.4 the results for the temperature T_0 and the Tsallis parameter q for all nine centrality classes obtained from fitting the Pb-Pb data at a beam energy of 5.02 TeV. Here we notice that the temperature varies from 98 MeV for the most central events and to 77 MeV for the most peripheral events, while the q parameter marginally rises from 1.14 to 1.15.

The resulting fits to the experimental data obtained in Pb-Pb collisions at $\sqrt{s_{NN}}$ = 5.02 TeV are shown in Figure 6.4 on the next page where again we follow the centrality classification introduced in [85]. As can be seen in Figure 6.4 the fits are very good for peripheral events and at low p_T, gradually worsening for the more central events where the fits at first overshoot the data above p_T values of about 3 GeV then rejoin the data and at larger values of p_T above about 20 GeV are below the data; a trend similar to what we observe at a beam energy of 2.76 TeV for Pb-Pb collisions.

TABLE 6.4: Values of q, T_0 and χ^2/NDF obtained using eq. (2.24) to fit charged hadron transverse momentum spectra measured by the ALICE collaboration in Pb-Pb collisions at $\sqrt{s_{NN}} = 5.02$ TeV [85].

Centrality Class		q	T_0 (MeV)	χ^2/NDF
1	(0-5)%	1.1405 ± 0.0009	98.2 ± 1.3	163.8/58
2	(5-10)%	1.1413 ± 0.0009	97.8 ± 1.4	154.1/58
3	(10-20)%	1.1424 ± 0.0009	96.8 ± 1.3	142.7/58
4	(20-30)%	1.1438 ± 0.0009	94.8 ± 1.2	126.6/58
5	(30-40)%	1.1449 ± 0.0009	92.5 ± 1.2	104.9/58
6	(40-50)%	1.1467 ± 0.0009	88.8 ± 1.2	86.17/58
7	(50-60)%	1.1478 ± 0.0009	85.3 ± 1.2	61.57/58
8	(60-70)%	1.1489 ± 0.0009	81.3 ± 1.2	37.62/58
9	(70-80)%	1.1503 ± 0.0010	77.4 ± 1.2	30.3/58

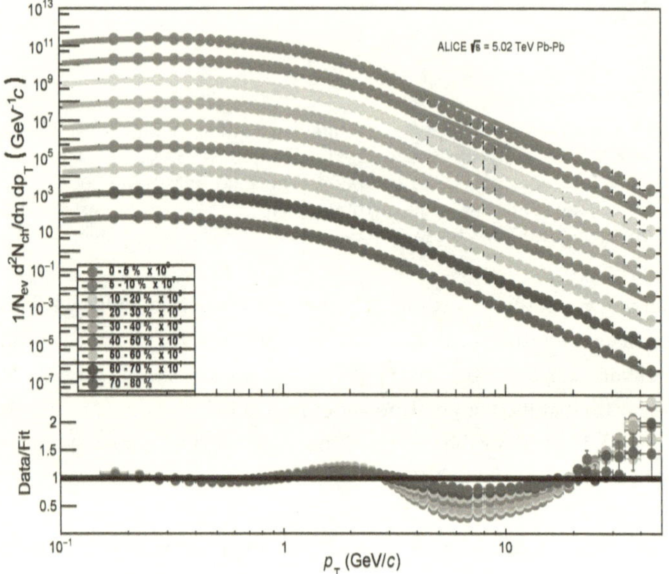

FIGURE 6.4: Transverse momentum distributions measured by the ALICE collaboration in Pb-Pb collisions at $\sqrt{s_{NN}} = 5.02$ TeV [85] as a function of centrality. The solid lines represent fits made using the Tsallis distribution eq. 2.23. The lower part of the figure shows the ratio of the data divided by the fit value.

Following the same procedure as before, we present in Table 6.5 the results for the temperature T_0 and the Tsallis parameter q for all nine centrality classes obtained from fitting the Xe-Xe data at a beam energy of 5.44 TeV. Here we notice that the temperature varies from 98 MeV for the most central events and to 79 MeV for the most peripheral events, while the q parameter marginally rises from 1.14 to 1.15.

TABLE 6.5: Values of q, T_0 and χ^2/NDF obtained from Tsallis fit using eq. (2.24) to fit charged hadron transverse momentum spectra measured by the ALICE collaboration in Xe-Xe collisions at $\sqrt{s_{NN}} = 5.44$ TeV [86].

Centrality Class		q	T_0 (MeV)	χ^2/NDF
1	(0-5)%	1.1423 ± 0.0011	97.0 ± 1.5	142.6/58
2	(5-10)%	1.1421 ± 0.0012	97.2 ± 1.7	117.8/58
3	(10-20)%	1.1432 ± 0.0011	95.6 ± 1.5	104/58
4	(20-30)%	1.1440 ± 0.0011	93.1 ± 1.4	86.59/58
5	(30-40)%	1.1461 ± 0.0011	90.2 ± 1.4	73.6/58
6	(40-50)%	1.1472 ± 0.0012	87.1 ± 1.4	61.02/58
7	(50-60)%	1.1478 ± 0.0013	84.7 ± 1.4	41.05/58
8	(60-70)%	1.1489 ± 0.0015	81.5 ± 1.6	32.62/58
9	(70-80)%	1.1488 ± 0.0016	78.9 ± 1.7	28.25/58

The resulting fits to the experimental data obtained in Xe-Xe collisions at $\sqrt{s_{NN}} = 5.44$ TeV are shown in Figure 6.4 on the next page where again we follow the centrality classification introduced in [86]. As can be seen in Figure 6.5 the fits are very good for peripheral events and at low p_T, gradually worsening for the more central events where the fits at first overshoot the data above p_T values of about 3 GeV then rejoin the data and at larger values of p_T above about 20 GeV are below the data; a trend similar to what we observe at a beam energy of 2.76 and 5.02 TeV for Pb-Pb collisions.

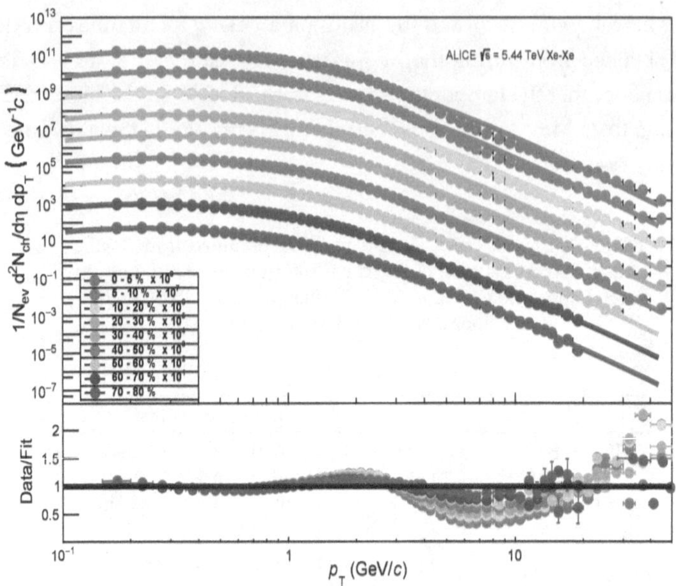

FIGURE 6.5: Transverse momentum distributions measured by the ALICE collaboration in Xe-Xe collisions at $\sqrt{s_{NN}}$ = 5.44 TeV [86] as a function of centrality. The solid lines represent fits made using the Tsallis distribution eq. (2.23). The lower part of the figure shows the ratio of the data divided by the fit value.

For all the three different collision energies, we observe same behaviour. The transverse momentum distributions tend to show an S shape for central collisions, this shape is difficult to reproduce using the Tsallis parameterization which has only two variables T_0 and q and the overall normalization fixed by the volume V. Clearly one more parameter would be needed to reproduce the shape for the most central events.

For the most central Pb-Pb collisions, a large number of nuclei is involved in the collision in comparison to the peripheral collisions, consequently, the system so created is a function of the colliding nuclei. Thus, peripheral collisions with very few participating nuclei can be thought of as being similar to a system created in $p - p$ collisions.

Our results are consistent with those obtained in analyses using the blast-wave [87] formalism [88, 89, 90] but they are considerably lower than those obtained recently in [91, 92] also the dependence in centrality is reversed. As usual, the Tsallis parameter q can be determined with an excellent accuracy.

6.4   Thermodynamic variables

Having deduced the temperature T_0 and the Tsallis parameter q at kinetic freeze-out from the transverse momentum distributions for three beam energies, we now proceed calculating the energy density, pressure, particle number density and entropy given by eq. (2.27) to eq. (2.32) respectively.

6.4.1   Energy density at kinetic freeze-out

In Table 6.6 we collect all the results obtained for the energy density of charged hadrons as a function of centrality for the three different energies compared with a few other energy densities. The entry for the chemical freeze-out energy density has been obtained using the latest version of THERMUS [16] [1]. The latter has been calculated from all hadronic resonances and is not limited to the charged particles only. It has been shown recently that the chemical freeze-out temperature is approximately independent of centrality [88, 93, 94].

For comparison we also show the energy density inside a proton calculated using the charge radius of the proton given as 0.875 fm and the mass of the proton as listed in the particle data booklet PDG [95]. The difference between the kinetic and chemical freeze-out results is not surprising in view of the fact that the energy density changes as T^4 for massless particles. Also show in Table 6.6 is the energy density obtained in the phase transition region obtained using Lattice QCD [96].

In Figure 6.6 we show the energy density divided by the kinetic freeze-out temperature to the fourth power so as to have a dimensionless quantity. As can be seen in this figure, the dependence on the centrality class is strongly reduced. It is also not unexpected that the ϵ/T^4 values are slightly higher at a beam energy of 5.02 TeV than at 2.76 TeV. Here we also notice that the values for the Xe-Xe collisions are similar to those for Pb-Pb collisions results at 5.02 TeV.

---

[1]B. Hippolyte and Y. Schutz, https://github.com/thermus-project/THERMUS

TABLE 6.6: Values for the energy density of charged hadrons, expressed in GeV fm^{-3} obtained using eq. (2.27) for the different centrality classes in Pb-Pb and Xe-Xe collisions. The energy density at chemical freeze-out has been calculated at $T = 153 \pm 3.18$ MeV for the most central Pb-Pb collisions as given in [16].

Centrality Class	ϵ at 2.76 TeV	ϵ at 5.02 TeV	ϵ at 5.44 TeV
1 (0-5)%	0.03272 ± 0.00041	0.03933 ± 0.00049	0.03782 ± 0.0006
2 (5-10)%	0.03218 ± 0.00041	0.03860 ± 0.00049	0.03814 ± 0.0006
3 (10-20)%	0.03153 ± 0.00039	0.03732 ± 0.00047	0.03570 ± 0.0005
4 (20-30)%	0.02938 ± 0.00037	0.03487 ± 0.00045	0.03192 ± 0.0005
5 (30-40)%	0.02696 ± 0.00035	0.03148 ± 0.00042	0.02837 ± 0.0004
6 (40-50)%	0.02339 ± 0.00032	0.02669 ± 0.00036	0.02444 ± 0.0004
7 (50-60)%	0.01964 ± 0.00028	0.02241 ± 0.00031	0.02153 ± 0.0004
8 (60-70)%	0.01604 ± 0.00025	0.01809 ± 0.00026	0.01824 ± 0.0003
9 (70-80)%	0.01356 ± 0.00024	0.01458 ± 0.00023	0.01563 ± 0.0003
Proton [95]	0.334		
Chemical freeze-out [16]	0.3625 ± 0.0716		
Lattice QCD [96]	0.34 ± 0.16		
Cold nuclear matter	0.16		

An estimate can now be made of the lifetime of the hadronic stage between chemical freeze-out and the kinetic freeze-out using the Bjorken model [62] with isentropic expansion which gives:

$$\epsilon(\tau) = \epsilon(\tau_0) \left(\frac{\tau_0}{\tau}\right)^{4/3}. \tag{6.1}$$

For the top 5% central Pb-Pb collisions at 5.02 TeV, this leads to

$$\frac{\tau(\text{kinetic fo})}{\tau(\text{chemical fo})} = \left(\frac{\epsilon(\text{chemical})}{\epsilon(\text{kinetic})}\right)^{3/4} \approx 3.9, \tag{6.2}$$

where the energy density at kinetic freeze-out has been corrected by a factor 3/2 to take into account the neutral hadrons. For example, if chemical freeze-out happens at $\tau = 10$ fm, then kinetic freeze-out happens at $\tau = 39$ fm. The chemical freeze-out time could be different for different centralities.

If the chemical freeze-out time is the same or at least similar for all centralities then one has to conclude that the time between chemical and kinetic freeze-out is longer for peripheral collisions than for central collisions. As a reminder, in the Bjorken model [62], which is an inside-outside cascade, the

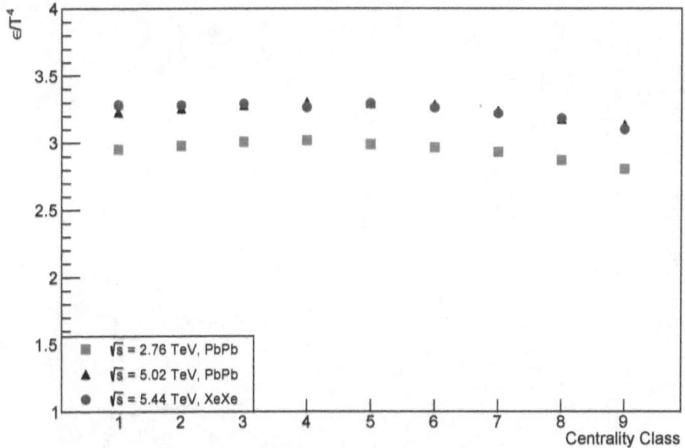

FIGURE 6.6: Energy density of charged hadrons divided by the kinetic freeze-out temperature in Pb-Pb collisions at 2.76 and 5.02 TeV [85] and for Xe-Xe collisions at 5.44 TeV [86] as a function of centrality class calculated using eq. (2.27).

central region freezes out first while the peripheral region remains hot. As this is a scaling model, there is no natural cut-off time.

6.4.2 Pressure at kinetic freeze-out

The pressure plays an important role in the hydrodynamic description of heavy-ion collisions, e.g. in the study of shock waves or the speed of sound in a hadronic gas. In the present analysis, the pressure can be determined explicitly from the following eq. (2.12). The results are shown in Figure 6.7 where one notices a clear, expected, increase in the pressure when going from peripheral collisions to central ones.

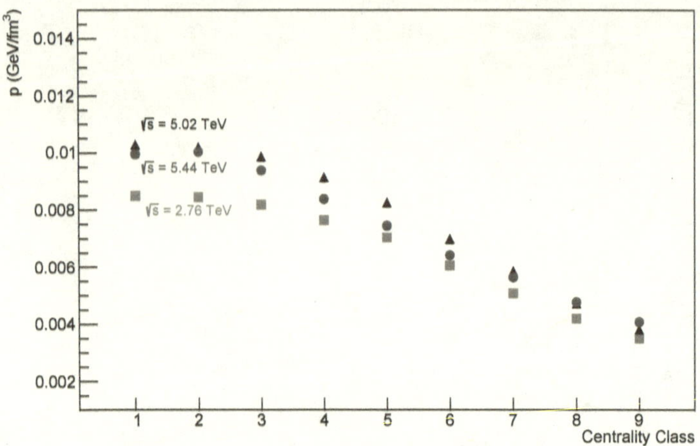

FIGURE 6.7: Pressure of charged hadrons at kinetic freeze-out in Pb-Pb collisions at 2.76 and 5.02 TeV [85] and for Xe-Xe collisions at 5.44 TeV [86] as a function of centrality class calculated using eq. (2.28).

For the results are shown in Figure 6.7 , we have also checked explicitly that the inequality:

$$\epsilon \geq 3P,\tag{6.3}$$

is always satisfied.

6.4.3 Entropy density at kinetic freeze-out

The entropy is an important quantity because it plays a major role in hydrodynamic expansion calculations where entropy is sometimes assumed to be conserved when going from the quark-gluon plasma phase to the hadronic phase. This is for example the case in the Bjorken model [62]. It is difficult to relate it directly to a measurable quantity and it is often indirectly linked to the particle number. The entropy density is given by eq. (2.32) where the parameters T and q are taken from Table 6.3 for Pb-Pb collisions at $\sqrt{s_{NN}} = 2.76$ TeV, Table 6.4 for collisions at 5.02 TeV and Table 6.5 for Xe-Xe collisions at $\sqrt{s_{NN}} = 5.44$ TeV. The results are shown in Figure 6.8 where the entropy density has been divided by T^3 so as to have a dimensionless quantity. There is also a small increase when the beam energy is increased from $\sqrt{s_{NN}} = 2.76$ to 5.02 TeV for Pb-Pb collisions.

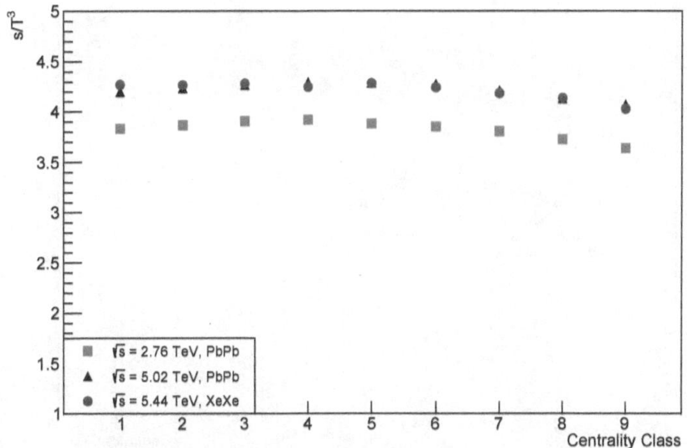

FIGURE 6.8: Entropy density of charged hadrons divided by the kinetic freeze-out temperature to the third power in Pb-Pb collisions at 2.76 and 5.02 TeV [85] and for Xe-Xe collisions at 5.44 TeV [86] as a function of centrality class calculated using eq. (2.32) .

In Figure 6.8, we have checked explicitly that the thermodynamic relation,

$$\epsilon + P = T_0 s,\tag{6.4}$$

holds. This is further confirmation of the consistency of having the chemical potential μ equal to zero for the collisions under consideration. As this is done at kinetic freeze-out and not at chemical freeze-out, this is a non-trivial observation. At chemical freeze-out the chemical potentials must be zero because of the equal numbers of particles and antiparticles. At thermal freeze-out however it is only required that the chemical potentials for particles and antiparticles be equal but not necessarily zero. It is still legitimate to have chemical potentials at kinetic freeze-out but they change the normalization and no longer determine relative abundances.

6.4.4 Particle density at kinetic freeze-out

For completeness we show the particle density calculated using eq. (2.29) in Figure 6.9. This is clearly well below the interior density of a heavy nucleus which is 0.17 nucleons/fm^3 [95].

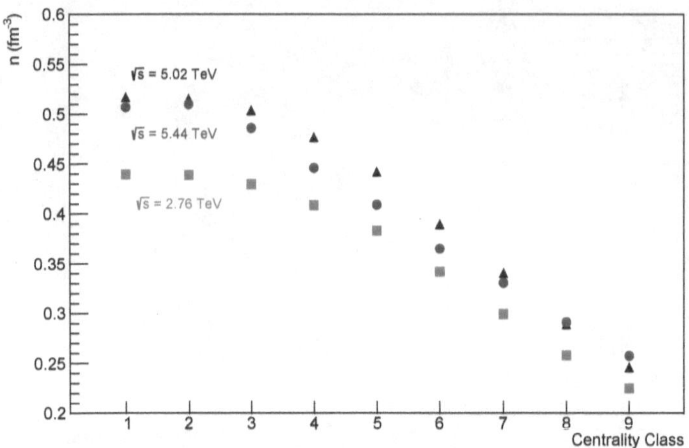

FIGURE 6.9: Charged particle density at kinetic freeze-out in Pb-Pb collisions at 2.76 and 5.02 TeV [85] and for Xe-Xe collisions at 5.44 TeV [86] as a function of centrality class calculated using eq. (2.29) .

6.5 Summary

The transverse momentum distributions for identified particles measured in the Pb-Pb collisions at $20, 30$ GeV [83], $40, 80$ and 158 GeV [84] with the NA 49 Collaboration, the primary charged particles measured in Pb-Pb collisions at $\sqrt{s_{NN}} = 2.76$ and 5.02 TeV by the ALICE collaboration [85] as well as for Xe-Xe collisions at 5.44 TeV [86] have been analysed using a thermodynamically consistent form of the Tsallis distribution based on eq. (2.6).

For the charged particles, the Tsallis distribution gives a very good description of the transverse momentum distributions for the most peripheral collisions, gradually worsening for the most central events where the fits at first overshoot the data at large values of p_T and in the end are below the data, which is a matter of further exploration.

The temperature T and the Tsallis parameter q have been determined at the three beam energies for all the centrality classes. Using the values obtained we then determined the energy density, ϵ, pressure, P, entropy density, s and the particle density, n at kinetic freeze-out as a function of the centrality classes. As expected, the values of all thermodynamic quantities

show an increase towards higher centrality class and at higher beam energies.

It is determined that in the final freeze-out stage, the energy density reaches
a value of about 0.039 GeV/fm^3 for the most central collisions at $\sqrt{s_{NN}} =$
5.02 TeV. This is less than that at chemical freeze-out where the energy density is about 0.36 GeV/fm^3. This decrease approximately follows a T^4 law. It
can be concluded that, together with the results obtained at chemical freeze-
out, the thermodynamic quantities presented here provide information about
the evolution of the thermodynamic quantities during the evolution of the
hadronic phase from chemical to kinetic freeze-out.

For the most central Pb-Pb collisions, a large number of nuclei is involved
in the collision in comparison to the peripheral collisions. Consequently, the
system so created is a function of the colliding nuclei. Thus, peripheral collisions with very few participating nuclei can be thought of as being similar to
a system created in $p - p$ collisions.

Part III

Applications of extensive statistics

Chapter 7

Basic literature

This chapter introduces the three ensemble classes which are implemented in the THERMUS package [16]. The literature presented this chapter has been presented in [3, 4].

7.1 Thermal model fits

The statistical model in the form of the hadron resonance gas model has been successful [97, 98] in describing the composition of the final state e.g. the yields of pions, kaons, protons and other hadrons. In these descriptions use is made of the grand canonical ensemble and the canonical ensemble with exact strangeness conservation.

The yields produced in heavy-ion collisions have been the subject of intense discussions over the past few years and several proposals have been made in view of the fact that the number of pions is underestimated while the number of protons is overestimated. Several proposals to improve on this have been made recently:

- Incomplete hadron spectrum [99],

- chemical non-equilibrium at freeze-out [100, 101, 102],

- modification of hadron abundances in the hadronic phase [103, 104],

- separate freeze-out for strange and non-strange hadrons [105, 106, 107],

- excluded volume interactions [108],

- energy dependent Breit-Wigner widths [109],

- use the phase shift analysis to take into account repulsive and attractive interactions [110, 111],

- use the K-matrix formalism to take interactions into account [112].

These proposals improve the agreement with the observed yields and furthermore, some of them change the chemical freeze-out temperature, T_{ch} in only a minimal way like those presented recently in [109, 111]. In the present analysis we therefore kept to the basic structure of the thermal model with a single freeze-out temperature and focus on the resulting thermal parameters, namely, the chemical freeze-out temperature T_{ch}, the strangeness saturation factor γ_s and the radius.

In our study, we consider in addition to the above, the use of the canonical ensemble with exact baryon, strangeness and charge conservation and conduct a systematic analysis of the dependence on the charged particle multiplicity $dN_{ch}/d\eta$ for the first time. A similar analysis was done in [113, 114, 115] for $p - p$ collisions at 200 GeV but without the dependence on charged multiplicity.

7.2   Model description

The identifying feature of the thermal model is that all the resonances listed in [116] are assumed to be in thermal and chemical equilibrium. This assumption drastically reduces the number of free parameters as this stage is determined by just a few thermodynamic variables namely, the chemical freeze-out temperature T_{ch}, the various chemical potentials μ determined by the conserved quantum numbers and by the volume V of the system [117, 118]. It has been shown that this description is also the correct one [119, 120, 121] for a scaling expansion as first discussed by Bjorken [62]. After integration over p_T these authors have shown that:

$$\frac{dN_i/dy}{dN_j/dy} = \frac{N_i^0}{N_j^0},$$

(7.1)

where N_i^0 is the particle yield as calculated in a fireball at rest. Hence, in the Bjorken model with longitudinal scaling and radial expansion, the effects of hydrodynamic flow cancel out in ratios provided the temperature is the same on the freeze-out surface.

The THERMUS [122] package treats the final state composition of a fireball resulting from high energy collision as an ideal gas of hadrons and resonances and all the particle species are assumed to be in thermal equilibrium. When utilising the thermal model, one can make a choice of the ensemble with which to treat the quantum numbers, namely B (baryon number), S (strangeness) and Q (charge), thus, the chemical potential for particle species i is given by,

$$\mu_i = B_i \mu_B + S_i \mu_S + Q_i \mu_Q. \tag{7.2}$$

In the <u>Grand canonical ensemble</u> (GCE): the conservation of quantum numbers is implemented using chemical potentials such that the quantum numbers are conserved on the average. The partition function depends on thermodynamic quantities and the Hamiltonian describing the system of N hadrons:

$$Z_{GCE} = \mathrm{Tr}\left[e^{-(H-\mu N)/T}\right] \tag{7.3}$$

which, in the framework of the thermal model considered here, leads to

$$\ln Z_{GCE}(T, \mu, V) = \sum_i g_i V \int \frac{d^3 p}{(2\pi)^3} \exp\left(-\frac{E_i - \mu_i}{T}\right) \tag{7.4}$$

in the Boltzmann approximation. The yield is given by:

$$N_i^{GCE} = g_i V \int \frac{d^3 p}{(2\pi)^3} \exp\left(-\frac{E_i - \mu_i}{T}\right). \tag{7.5}$$

We have put the chemical potentials equal to zero, as relevant for the beam energies considered here. The decays of resonances have to be added to the final yield

$$N_i^{GCE}(\text{total}) = N_i^{GCE} + \sum_j Br(j \to i) N_i^{GCE}. \tag{7.6}$$

In the <u>Strangeness canonical ensemble</u> (SCE): there are chemical potentials for baryon number B and charge Q but not for strangeness which is fixed exactly:

$$Z_{SCE} = \mathrm{Tr}\left[e^{-(H-\mu N)/T}\delta_{(S,\sum_i S_i)}\right] \tag{7.7}$$

The delta function imposes exact strangeness conservation, requiring overall strangeness to be fixed to the value S, in this paper we will only consider the case where overall strangeness is zero, $S = 0$. This change leads to [123]:

$$Z_{SCE} = \frac{1}{(2\pi)} \int_0^{2\pi} d\phi e^{-iS\phi} Z_{GCE}(T, \mu_B, \lambda_S) \tag{7.8}$$

where the fugacity factor is replaced by

$$\lambda_S = e^{i\phi} \tag{7.9}$$

$$N_i^{SCE} = V \frac{Z_i^1}{Z_{S=0}^C} \sum_{k,p=-\infty}^{\infty} a_3^p a_2^k a_1^{-2k-3p-s} I_k(x_2) I_p(x_3) I_{-2k-3p-s}(x_1),$$

$$(7.10)$$

where $Z_{S=0}^C$ is the canonical partition function

$$Z_{S=0}^C = e^{S_0} \sum_{k,p=-\infty}^{\infty} a_3^p a_2^k a_1^{-2k-3p} I_k(x_2) I_p(x_3) I_{-2k-3p}(x_1),$$

where Z_i^1 is the one-particle partition function calculated for $\mu_S = 0$ in the Boltzmann approximation. The arguments of the Bessel functions $I_s(x)$ and the parameters a_i are introduced as,

$$a_s = \sqrt{S_s/S_{-s}} \ , \quad x_s = 2V\sqrt{S_s S_{-s}}, \tag{7.11}$$

where S_s is the sum of all $Z_k^1(\mu_S = 0)$ for particle species k carrying strangeness s. As previously, the decays of resonances have to be added to the final yield

$$N_i^{SCE}(\text{total}) = N_i^{SCE} + \sum_j Br(j \rightarrow i) N_i^{SCE}. \tag{7.12}$$

In the <u>Full canonical ensemble</u> (FCE): all charges are fixed exactly and there are no chemical potentials. This ensemble can be called the canonical ensemble with exact implementation of B, S and Q conservation. The partition function is given by:

$$Z_{FCE} = \text{Tr}\left[e^{-(H-\mu N)/T} \delta_{(B,\Sigma_i B_i)} \delta_{(Q,\Sigma_i Q_i)} \delta_{(S,\Sigma_i S_i)} \right] \tag{7.13}$$

$$Z_{FCE} = \frac{1}{(2\pi)^3} \int_0^{2\pi} d\psi e^{-iB\alpha} \int_0^{2\pi} d\phi e^{-iQ\psi} \int_0^{2\pi} d\alpha e^{-iS\phi} Z_{GCE}(T, \lambda_B, \lambda_Q, \lambda_S)$$

$$(7.14)$$

where the fugacity factors have been replaced by

$$\lambda_B = e^{i\alpha}, \quad \lambda_Q = e^{i\psi}, \quad \lambda_S = e^{i\phi}. \tag{7.15}$$

As before, the decays of resonances have to be added to the final yield

$$N_i^{FCE}(\text{total}) = N_i^{FCE} + \sum_j Br(j \rightarrow i) N_i^{FCE}. \tag{7.16}$$

In this case the analytic expression becomes very lengthy and we refrain from writing it down here, it is implemented in the THERMUS program [16]. In all three cases we have also taken into account the strangeness saturation factor γ_s [124] which enters as a multiplicative factor, raised to the power of the strangeness content, in the particle yields.

7.3 Summary

In this chapter, we have described the three ensemble classes which will inform the data analysis presented in the next chapter.

Chapter 8

Thermal model fits

8.1 Introduction

This chapter presents a systematic analysis of the dependence of T_{ch}, γ_s and the volume on the charged particle multiplicity for the three ensembles namely: the grand canonical ensemble (Grand Canonical), the canonical ensemble with exact strangeness conservation (Canonical S) and the Canonical ensemble with exact implementation of B, S and Q conservation (Canonical BSQ).

These are applied to $p-p$ collisions at 7 TeV [125], $p-$Pb collisions at 5.02 TeV [126, 127] and to Pb-Pb collisions at 2.76 TeV [128, 129, 130] in the central region of rapidity. A similar analysis was done in [113, 114, 115] for $p-p$ collisions at 200 GeV but without the dependence on charged multiplicity

It is well known that in this kinematic region, one has particle - antiparticle symmetry and therefore there is no net baryon density and also no net strangeness. It therefore follows that the different ensembles should nevertheless give different results because of the way they are implemented. The results in this chapter were presented in [3, 4].

8.1.1 Applications of extensive statistics

In this study, the picture of a heavy-ion collision proposed is as follows: that the chemical freeze-out happens with Boltzmann-Gibbs statistics leading to a consistent picture of the hadronic yields, this leads to the deduction that the yields calculated using the Thermal model will model differ simply because of the way in which each ensemble is implemented.

1. **Comparison of $p-p$, $p-$Pb and Pb-Pb collisions in the thermal model**

   We present a systematic analysis of the dependence of T_{ch}, γ_s and the volume on the charged particle multiplicity $dN_{ch}/d\eta$. A similar analysis was done in [113, 114, 115] for $p - p$ collisions at 200 GeV but without the dependence on charged multiplicity.

We will show that the difference between the ensembles used disappears if the final state multiplicity is large. All calculations were done using THERMUS [122]. This model is customised for implementation in an Object Oriented Framework for large data analysis (ROOT) [64], and the results are plotted using the Grace [131] plotting software tool.

8.2   Method

For $p - p$ collisions we have considered the five particle species listed in Table 8.1 where we also compare the measured values with the model calculations for the different ensembles. For $p-$Pb and Pb-Pb collisions we included the Ω measurements in our analysis, such that six particle species were considered for $p-$Pb and Pb-Pb.

We have checked explicitly that for the five multiplicity bins in $p - p$ collisions where the Ω has also been measured, there is no difference in the outcome for the values of T_{ch}, γ_s and the radius. The ϕ meson is not described very well as shown in [132] and has not been included in our calculations. All our calculations were done using the latest version of THERMUS [16] [1].

8.3   Comparison of different ensembles

In Table 8.1, we present a comparison of the dN/dy values as calculated using the three different ensembles for the most central multiplicity events. The results from THERMUS are comparable to the data.

In Figure 8.1 we show the chemical freeze-out temperature as a function of the multiplicity of hadrons in the final state [125]. As explained in the previous section the freeze-out temperature has been calculated using three different ensembles. The highest values are obtained using the canonical ensemble with exact conservation of three quantum numbers, baryon number B, strangeness S and charge Q, all of them being set to zero as is appropriate for the central rapidity region in $p - p$ collisions at 7 TeV. In this ensemble the temperature drops strongly from the lowest to the highest multiplicity.

---

[1] B. Hippolyte and Y. Schutz, https://github.com/thermus-project/THERMUS

TABLE 8.1: Comparison between measured and fitted values
for $p - p$ collisions at 7 TeV for V0M multiplicity class II.

Particle Species	dN/dy (data)	dN/dy (model)		
		Canonical S	Canonical B, S, Q	Grand Canonical
$\pi^{\pm}$	7.88 ± 0.38	6.78	6.76	6.96
K^0_S	1.04 ± 0.05	1.16	1.16	1.15
$p, (\bar{p})$	0.44 ± 0.03	0.50	0.50	0.50
Λ	0.302 ± 0.020	0.259	0.262	0.246
$\Xi^-(\bar{\Xi}^+)$	0.0358 ± 0.0023	0.035	0.035	0.036

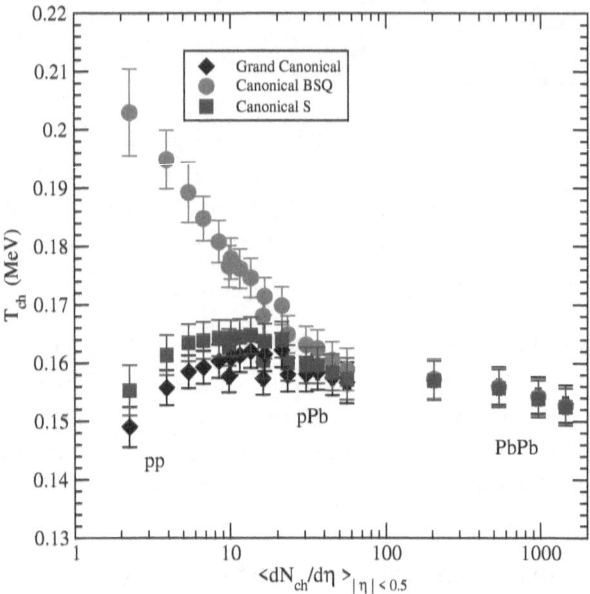

FIGURE 8.1: The chemical freeze-out temperature, T_{ch}, obtained for three different ensembles. The black diamonds represent results obtained using the grand canonical ensemble, the blue squares are for the exact strangeness conservation while the red circles are the results with the built-in exact baryon number, strangeness and charge conservation.

The lowest values for T_{ch} are obtained when using the grand canonical ensemble, in this case the conserved quantum numbers are again zero. The results are clearly different from those obtained in the previous ensemble, especially in the low multiplicity intervals. They gradually approach each other and they become equal at the highest multiplicities.

For comparison with the previous two cases we also calculated T_{ch} using the canonical ensemble with only strangeness S being exactly conserved using the method presented in [123]. In this case the results are close to those obtained in the grand canonical ensemble, with the values of T_{ch} always slightly higher than in the grand canonical ensemble. Again for the highest multiplicity interval the results become equivalent.

At very low multiplicities (corresponding to $p - p$ collisions), the canonical BSQ ensemble leads to results which are incompatible with those obtained from Lattice Gauge Theory [133] which indicate that hadrons cannot exist above the critical temperature which has been estimated to be about 156.5 ± 1.5 MeV. As can be seen in Figure 8.1, even though all the ensembles produce different results, the general trend is that for high multiplicities the results converge to a common value which is close to 160 MeV.

In Figure 8.2 we show results for the strangeness saturation factor γ_s [124] as a function of the multiplicity of hadrons. In this case we obtain some quite substantial differences in each one of the three ensembles considered. The highest values being found in the canonical ensemble with exact strangeness conservation. As the multiplicity increases, the values of γ_s become compatible with unity, i.e. with chemical equilibrium for all light flavours.

The Canonical S and Canonical BSQ do not give similar result (or trend) for $p - p$ collisions, this brings into question the suitability of using these two prescriptions for $p - p$ collisions. The two however give similar trend for $p-$Pb and Pb-Pb collisions.

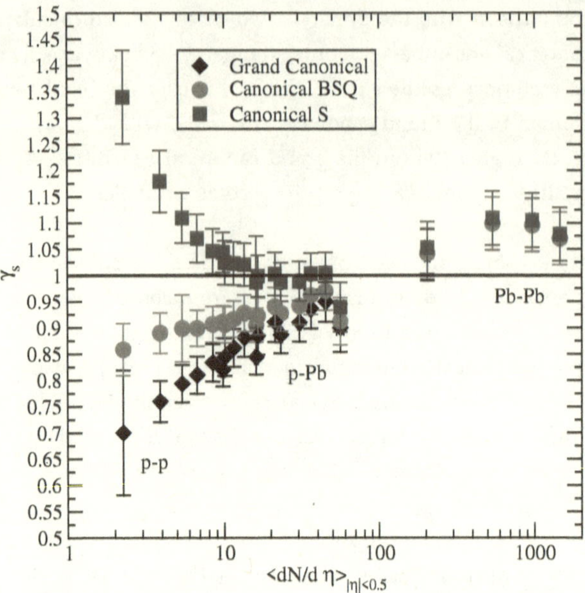

FIGURE 8.2: The strangeness saturation factor γ_s obtained for three different ensembles. The black diamonds represent results obtained using the grand canonical ensemble, the blue squares are for the exact strangeness conservation while the red circles are the results with the built-in exact baryon number, strangeness and charge conservation.

In Figure. 8.3 the volume at chemical freeze-out obtained in the three ensembles is presented as a function of the multiplicity of hadrons. As in the previous figures, the results become independent of the ensemble chosen for the highest multiplicities while showing clear differences for low multiplicities.

Again, in Figure. 8.3, a plot of the volume clearly shows that the number of particles is linearly proportional to volume in agreement with eq. (7.5).

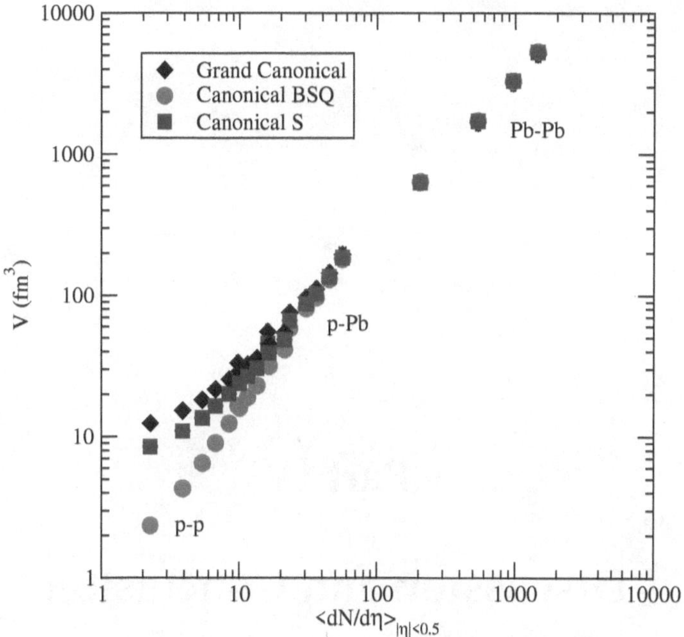

FIGURE 8.3: The chemical freeze-out volume obtained for three different ensembles. The black diamonds represent results obtained using the grand canonical ensemble, the blue squares are for the exact strangeness conservation while the red circles are the results with the built-in exact baryon number, strangeness and charge conservation.

8.4   Summary

The Thermal Model fit results on the dependence of T_{ch}, γ_s and the volume on the charged particle multiplicity have been presented for the to $p - p$ collisions at 7 TeV, $p-$Pb collisions at 5.02 TeV and to Pb-Pb collisions at 2.76 TeV in the central region of rapidity.

The results show that there is a strong correlation between some of the parameters. The very high temperature obtained in the canonical BSQ ensemble correlates with the small radius in the same ensemble. Particle yields increase with temperature but a small volume decreases them, hence the correlation between the parameters.

Part IV

Discussion and Conclusion

Chapter 9

Discussion and Conclusion

9.1 Applications of non-extensive statistics

We have successfully utilised non-extensive statistics in the analysis of transverse momentum spectra. The fits obtained with the Tsallis distribution has provided a very good description of the transverse momentum spectra in $p - p$, Pb-Pb and Xe-Xe collisions. The main result is that the parameters T_0 and q depend on the particle species and are different for pions, kaons and protons.

9.1.1 Fits to $p - p$, and Pb-Pb collisions

Overall, the conclusion we reach is that for the π's, K's, and protons the fit parameters are different and no full parameter universality exists. Thus, even though the Tsallis distribution provides a very good description of the transverse momenta distributions, it has to be concluded that the parameters are in fact different. It is worth noticing that the decays of resonances have not been taken into account and this could modify the conclusions presented here.

Adding flow to the Tsallis distribution reproduces the transverse momentum spectra in $p - p$, Pb-Pb and Xe-Xe collisions albeit no significant improvement to the quality of the fits. The main result is promising for pions, however, the extracted flow values are unrealistic for protons and kaons. This suggests that our formulation needs to be improved.

Taking note of the previous work by [37] which was only limited to pions and quarkonia only, we would have reached the same conclusion had we restricted our analysis to pions only; as such, we will not draw any conclusions from the fit results presented in this book and conclude that the inclusion of flow in the Tsallis prescription remains an open question.

The proposed inclusion of the chemical potential as one of the parameters to the Tsallis distribution has been successful: from the comparison of temperature values at zero chemical potential to the equivalent which is deducted from the fits at non-zero chemical potential. This result confirms that the variables T, V, q, μ in the Tsallis distribution have a redundancy for $\mu \neq 0$ as previously suggested by [47].

9.1.2   Determining thermodynamic parameters

For the charged particles, the Tsallis distribution also gives a very good description of the transverse momentum distributions for the most peripheral collisions, gradually worsening for the most central events where the fits at first overshoot the data at large values of p_T and in the end are below the data, which is a matter of further exploration.

The temperature T_0 and the Tsallis parameter q have been determined at the three beam energies for all the centrality classes. Using the values obtained we then determined the energy density, ϵ, pressure, P, entropy density, s and the particle density, n at kinetic freeze-out as a function of the centrality classes. As expected, the values of all thermodynamic quantities show an increase towards higher centrality class and at higher beam energies.

It is determined that in the final freeze-out stage, the energy density reaches a value of about 0.039 GeV/fm^3 for the most central collisions at $\sqrt{s_{NN}} =$ 5.02 TeV. This is less than that at chemical freeze-out where the energy density is about 0.36 GeV/fm^3. This decrease approximately follows a T^4 law. It can be concluded that, together with the results obtained at chemical freeze-out, the thermodynamic quantities presented here provide information about the evolution of the thermodynamic quantities during the evolution of the hadronic phase from chemical to kinetic freeze-out.

9.2 Applications of extensive statistics

We have investigated three different ensembles to analyse the variation of particle yields with the multiplicity of charged particles produced in proton-proton collisions at the center-of-mass energy of $\sqrt{s} = 7$ TeV [125], $p-$Pb collisions at 5.02 TeV [126, 127] and Pb-Pb collisions at 2.76 TeV [128, 129, 130].

We have kept the basic structure of the thermal model as presented in [16] and focused on the resulting thermal parameters T_{ch}, γ_s and the volume and their dependence on the final state multiplicity. We note in this regards that recent improvements on the treatment of the particle yields do not lead to substantial changes of the chemical freeze-out temperature, T_{ch} [109, 111].

We take note of the following interesting new features:

- The grand canonical ensemble, the ensemble with strict strangeness conservation and the one with strict baryon number, strangeness and charge conservation agree very well for the particle composition in Pb-Pb collisions, they also agree well for $p-$Pb collisions but marked differences for p-p collisions are present. These differences disappear as the multiplicity of charged particles increases in the final state. Thus, $p - p$ collisions with high multiplicities agree with what is seen in large systems like $p-$Pb and Pb-Pb collisions. Quantitatively this agreement starts when there are at least 20 charged hadrons in the mid-rapidity interval being considered. It also throws doubt on the applicability of the thermal model as applied to $p - p$ collisions with low multiplicity.

- The convergence of the results in the three ensembles lends support to the idea that one reaches a thermodynamic limit where the results are independent of the ensemble being used.

We believe that it is of interest to note that all three ensembles lead to the same results when the multiplicity of charged particles $dN_{ch}/d\eta$ exceeds 20 at mid-rapidity. This could be interpreted as reaching the thermodynamic limit since the three ensembles lead to the same results. It would be of interest to extend this analysis to higher beam energies and higher multiplicity intervals.

The Thermal model fit results on the dependence of T_{ch}, γ_s and the volume on the charged particle multiplicity have been presented for the to $p - p$ collisions at 7 TeV, $p-$Pb collisions at 5.02 TeV and to Pb-Pb collisions at 2.76

TeV in the central region of rapidity.

The results show that there is a strong correlation between some of the parameters. The very high temperature obtained in the canonical BSQ ensemble correlates with the small volume in the same ensemble. Particle yields increase with temperature but a small volume decreases them, hence the correlation between the parameters.

Appendix A

Kinematic variables

This section is included for completeness and to introduce terms commonly used in this book, this section is based on the lecture notes given by [134]. We present a brief introduction to kinematic variables.

The restriction that the speed of light c is the same in all frames of reference leads to special relativity. In special relativity, the Lorentz transformations are given by

$$
\begin{aligned}
x' &= \gamma(x - vt) \\
t' &= \gamma\left(t - \frac{vx}{c^2}\right),
\end{aligned}
\tag{A.1}
$$

where the primed frame is viewed moving with speed v along the positive x-axis by the unprimed frame. The gamma factor appearing in eq. (A.1) is defined as

$$
\gamma \equiv \frac{1}{\sqrt{1 - v^2/c^2}}.
\tag{A.2}
$$

The momentum four-vector of a particle is defined as

$$
p^\mu = (E/c, p_x, p_y, p_z),
\tag{A.3}
$$

where beam direction is taken to be along the z-direction. The spatial components of the momentum vector are: p_z parallel to z-axis and the transverse components is p_x and p_y. The rapidity variable (y) is defined as

$$
y \equiv \frac{1}{2}\ln\frac{E + p_z}{E - p_z}.
\tag{A.4}
$$

The energy E and momentum p_z can be written in terms of transverse mass and rapidity as

$$
\begin{aligned}
E &= m_T \cosh y \\
p_z &= m_T \sinh y,
\end{aligned}
\tag{A.5}
$$

which we now substitute into the momentum four-vector eq. (A.3) to give

$$p^\mu \equiv (m_T \cosh y, p_x, p_y, m_T \sinh y).$$

If the mass of the particle is not known, one may opt for the pseudo-rapidity (η) variable defined by:

$$\begin{aligned}
\eta &\equiv \frac{1}{2} \ln \frac{p + p_z}{p - p_z} \\
&= \frac{1}{2} \ln \frac{p(1 + \cos\theta)}{p(1 - \cos\theta)} \\
\eta &= -\ln \tan(\theta/2),
\end{aligned} \qquad (A.6)$$

instead of the usual rapidity variable. With pseudo-rapidity, we only need to know the scattering angle θ; which is an angle at which a particle is emitted with respect to the beam axis. To change from rapidity to pseudo-rapidity, one uses the following relation [135]:

$$\frac{d^2 N}{dp_T d\eta} = \sqrt{1 - \frac{m^2}{m_T^2 \cosh^2 y}} \frac{dN}{dp_T dy}, \qquad (A.7)$$

where m is the rest mass of the respective particle. At mid -rapidity, $y = 0$, this implies that eq. (A.7) simplifies to

$$\frac{d^2 N}{dp_T d\eta} = \frac{p_T}{m_T} \frac{dN}{dp_T dy}, \qquad (A.8)$$

which introduces an extra factor of p_T/m_T, one needs to take care of this factor when fitting the transverse momentum spectra data with the Tsallis distribution.

Appendix B

Interpretation of the q parameter

Starting from the most commonly used distribution [35]

$$G_q(x) = C_q \left(1 - (q-1)\frac{x}{\lambda}\right)^{\frac{1}{1-q}},$$ (B.1)

which tends to

$$G_{q\to 1}(x) = g(x) = c\exp\left(-\frac{x}{\lambda}\right),$$ (B.2)

as q tends to 1 as previously stated. For values of $q > 1$, the parameter q represents fluctuations present in eq. $(B.1)$. To prove this, we analyse the influence of fluctuations of the parameter $\frac{1}{\lambda}$ present in eq. $(B.2)$ on the final result by deducing a function $f(\frac{1}{\lambda})$ which leads from eq. $(B.2)$, a pure exponential function to eq. $(B.1)$ a powerlike distribution and also describes fluctuations about the mean value $1/\lambda_0$ such that [35]

$$G_q(x;\lambda_0) = C_q \left(1 + \frac{x}{\lambda_0}\frac{1}{\alpha}\right)^{-a},$$ (B.3)

where an abbreviation $\alpha = \frac{1}{q-1}$ was introduced. Recalling from the Euler gamma function [35]

$$\left(1 + \frac{x}{\lambda_0}\frac{1}{\alpha}\right)^{-a} = \frac{1}{\Gamma(\alpha)}\int_0^\infty d\xi\, \xi^{\alpha-1} \times \exp\left(-\xi\left(1 + \frac{x}{\lambda_0}\frac{1}{\alpha}\right)\right).$$ (B.4)

Using a change of variables $\frac{\xi}{\lambda_0}\frac{1}{\alpha} = \frac{1}{\lambda} \implies \xi = \frac{\lambda_0\alpha}{\lambda}$ and $d\xi = \lambda_0\alpha\, d(\frac{1}{\lambda})$. The argument inside the exponent simplifies to $-\frac{\lambda_0\alpha}{\lambda} - \frac{x}{\lambda}$ after simplification.

Substituting all these into eq. $(B.4)$ gives

$$\left(1+\frac{x}{\lambda_0}\frac{1}{\alpha}\right)^{-a} = \frac{1}{\Gamma(\alpha)}\int_0^\infty \lambda_0\alpha d\left(\frac{1}{\lambda}\right)\left(\frac{\lambda_0\alpha}{\lambda}\right)^{\alpha-1}$$
$$\times \exp\left(-\frac{\lambda_0\alpha}{\lambda}-\frac{x}{\lambda}\right).$$
$$= \frac{1}{\Gamma(\alpha)}\int_0^\infty \lambda_0\alpha d\left(\frac{1}{\lambda}\right)\left(\frac{\lambda_0\alpha}{\lambda}\right)^{\alpha-1}$$
$$\times \exp\left(-\frac{\lambda_0\alpha}{\lambda}\right)\exp\left(-\frac{x}{\lambda}\right), \tag{B.5}$$

from which we can write eq. $(B.3)$

$$G_q(x;\lambda_0) = C_q\int_0^\infty \frac{1}{\Gamma(\alpha)}\lambda_0\alpha d\left(\frac{1}{\lambda}\right)\left(\frac{\lambda_0\alpha}{\lambda}\right)^{\alpha-1}$$
$$\times \exp\left(-\frac{\lambda_0\alpha}{\lambda}\right)\exp\left(-\frac{x}{\lambda}\right),$$
$$= C_q\int_0^\infty \exp\left(-\frac{x}{\lambda}\right)f\left(\frac{1}{\lambda}\right)d\left(\frac{1}{\lambda}\right), \tag{B.6}$$

where the function $f\left(\frac{1}{\lambda}\right)$ is given by [35]

$$f\left(\frac{1}{\lambda}\right) = f_\alpha\left(\frac{1}{\lambda},\frac{1}{\lambda_0}\right)$$
$$= \frac{1}{\Gamma(\alpha)}(\alpha\lambda_0)\left(\frac{\lambda_0\alpha}{\lambda}\right)^{\alpha-1}\exp\left(-\frac{\alpha\lambda_0}{\lambda}\right). \tag{B.7}$$

Now that we have found the function $f(\frac{1}{\lambda})$, we proceed to finding the relative variance ω which is given by [35]

$$\omega = \frac{\left\langle\left(\frac{1}{\lambda}\right)^2\right\rangle - \left\langle\frac{1}{\lambda}\right\rangle^2}{\left\langle\frac{1}{\lambda}\right\rangle^2}, \tag{B.8}$$

where $\left\langle\frac{1}{\lambda}\right\rangle$ represents the mean value of $\frac{1}{\lambda}$. From the definition of a general gamma function, $\Gamma(\alpha) = \int_0^\infty e^{-t}t^{\alpha-1}dt$, we can take $t = \frac{\alpha\lambda_0}{\lambda} \implies dt = \alpha\lambda_0 d(\frac{1}{\lambda})$, with these, one can write out $\Gamma(\alpha)$ as

$$\Gamma(\alpha) = \int_0^\infty e^{\left(-\frac{\alpha\lambda_0}{\lambda}\right)}\left(\frac{\alpha\lambda_0}{\lambda}\right)^{\alpha-1}\alpha\lambda_0 d\left(\frac{1}{\lambda}\right). \tag{B.9}$$

The next step is to determine $\left\langle \frac{1}{\lambda} \right\rangle$ and $\left\langle \left(\frac{1}{\lambda} \right)^2 \right\rangle$ and substitute these into eq. $(B.8)$. The average $\left\langle \frac{1}{\lambda} \right\rangle$ is given by

$$
\begin{aligned}
\left\langle \frac{1}{\lambda} \right\rangle &= \int_0^\infty f\left(\frac{1}{\lambda}\right)\left(\frac{1}{\lambda}\right) d\left(\frac{1}{\lambda}\right) \\
&= \frac{1}{\Gamma(\alpha)} \int_0^\infty \exp\left(-\frac{\alpha\lambda_0}{\lambda}\right)\left(\frac{\lambda_0\alpha}{\lambda}\right)^{\alpha-1}\left(\frac{\alpha\lambda_0}{\lambda}\right) d\left(\frac{1}{\lambda}\right) \\
&= \frac{1}{\alpha\lambda_0\Gamma(\alpha)} \int_0^\infty \exp\left(-\frac{\alpha\lambda_0}{\lambda}\right)\left(\frac{\lambda_0\alpha}{\lambda}\right)^{\alpha} d\left(\frac{\alpha\lambda_0}{\lambda}\right) \\
&= \frac{\Gamma(\alpha+1)}{\alpha\lambda_0\Gamma(\alpha)} \\
&= \frac{\alpha\Gamma(\alpha)}{\alpha\lambda_0\Gamma(\alpha)} \\
&= \frac{1}{\lambda_0}.
\end{aligned}
\tag{B.10}
$$

The average $\left\langle \left(\frac{1}{\lambda} \right)^2 \right\rangle$ is given by

$$
\begin{aligned}
\left\langle \left(\frac{1}{\lambda}\right)^2 \right\rangle &= \int_0^\infty f\left(\frac{1}{\lambda}\right)\left(\frac{1}{\lambda}\right)^2 d\left(\frac{1}{\lambda}\right) \\
&= \frac{1}{\Gamma(\alpha)} \int_0^\infty \exp\left(-\frac{\alpha\lambda_0}{\lambda}\right)\left(\frac{\lambda_0\alpha}{\lambda}\right)^{\alpha-1}\left(\frac{\alpha\lambda_0}{\lambda^2}\right) d\left(\frac{1}{\lambda}\right) \\
&= \frac{1}{\alpha^2\lambda_0^2\Gamma(\alpha)} \int_0^\infty \exp\left(-\frac{\alpha\lambda_0}{\lambda}\right)\left(\frac{\lambda_0\alpha}{\lambda}\right)^{\alpha-1}\left(\frac{\alpha\lambda_0}{\lambda}\right)^2 d\left(\frac{\alpha\lambda_0}{\lambda}\right) \\
&= \frac{\Gamma(\alpha+2)}{\alpha^2\lambda_0^2\Gamma(\alpha)} \\
&= \frac{(\alpha+1)\alpha\Gamma(\alpha)}{\alpha\lambda_0\Gamma(\alpha)} \\
&= \frac{\alpha+1}{\alpha\lambda_0^2}.
\end{aligned}
\tag{B.11}
$$

Thus, the relative variance is now given by

$$
\omega = \frac{1}{\alpha} = q - 1,
\tag{B.12}
$$

which implies that the parameter q describes the relative variance of $\frac{1}{\lambda}$ [35]: this closes the proof.

Appendix A

Basic fitting macro

The following is a basic macro for fitting a transverse momentum spectra
with the Tsallis distribution.

```cpp
// pT spectrum of K+ obtained in pp@900 GeV with ALICE
#include "TMinuit.h"   // -----------------call the fitting routine
#include <fstream>
Double_t dndy,averagept;
Double_t hbarc = 0.197327 ;
Double_t pi=TMath::Pi();
Int_t i;
Double_t mass,qtsallis;
                        // -------------set the particle mass, q and g
Double_t m[3] = {0.13957018,0.493677,0.938272013};
Double_t q[3] = {1.148,1.175,1.158};
Double_t g[3] = {1,1,2};
                        // -------------------- define the fit function
Double_t Tsallis_Fit(const Double_t *x, const Double_t *p){
Double_t pt       = x[0];
Double_t temp     = p[0];
Double_t mu       = p[1];
Double_t radius   = p[2]/hbarc;
Double_t qtsallis= p[3];
Double_t mass     = p[4];
Double_t deg      = p[5];
Double_t mt       = TMath::Sqrt(pt * pt + mass*mass);
Double_t volume   = 4.0*TMath::Pi()*TMath::Power(radius,3.0)/3.0;
Double_t Fit_Fun = deg*mt*TMath::Power((1.0 + (qtsallis - 1.0)*((mt - mu)/temp
   -qtsallis/(qtsallis - 1.0));
Double_t value    = pt*volume*Fit_Fun/(TMath::Power(2.0*TMath::Pi(),2));
return value;}
```

```cpp
//        ---------- main macro starts here
void ALICE_pp_900GeV_all_loop(){
//      ---- ---------- set the plot canvas
TCanvas *c1 = new TCanvas("c1","positive",800,600);
c1->SetFillColor(0);
c1->SetLogy(1);
c1->SetLogx(0);
gStyle->SetOptTitle(0);
TGraphErrors *gr[3];
TF1 *fTsallis[3];
// ========= start loop for pi +, K+ and p =============
for(i=0;i<3;i++){
//      ------------- read the data
gr[i] = new TGraphErrors((char*)(Form("alice_900_%d.dat",i)));
//      ------------- call the fit function
fTsallis[i] = new TF1("fTsallis", Tsallis_Fit, 0.0, 3.0, 6);
fTsallis[i]->SetParNames("T", "mu", "R","q","mass","degeneracy");
if(i==0){        // --------- initialise parameters
fTsallis[0]->SetParameter(0, 0.0758);
fTsallis[0]->SetParError(0, 0.0001);
fTsallis[0]->SetParameter(1, 0.10);
fTsallis[0]->SetParError(1, 0.0001);
fTsallis[0]->SetParameter(2, 4.0);
fTsallis[0]->SetParError(2, 0.00001);}
if(i==1){
fTsallis[1]->SetParameter(0, 0.065);
fTsallis[1]->SetParError(0, 0.0001);
fTsallis[1]->SetParameter(1, 0.05);
fTsallis[1]->SetParError(1, 0.0001);
fTsallis[1]->SetParameter(2, 3.5);
fTsallis[1]->SetParError(2, 0.00001);}
if(i==2){
fTsallis[2]->SetParameter(0, 0.05822);
fTsallis[2]->SetParError(0, 0.000001);
fTsallis[2]->SetParLimits(0, 0.0001,0.3);
fTsallis[2]->SetParameter(1, 0.2416);
fTsallis[2]->SetParError(1, 0.00001);
fTsallis[2]->SetParameter(2, 3.1567);
fTsallis[2]->SetParError(2, 0.000001);}
```

```cpp
fTsallis[i]->FixParameter(3,q[i]);
fTsallis[i]->FixParameter(4,m[i]);
fTsallis[i]->FixParameter(5,g[i]);
fTsallis[i]->SetLineWidth(3);
fTsallis[i]->SetMarkerColor(600+32*i);
fTsallis[i]->SetLineColor(600+32*i);
if(i==2){fTsallis[i]->SetMarkerColor(1);
fTsallis[i]->SetLineColor(1);}
gr[i]->SetMinimum(1.0e-4);
gr[i]->SetMaximum(10.0);
gr[i]->GetXaxis()->SetRangeUser(0.0,3.0);
gr[i]->GetXaxis()->SetTitle("p_{T} (GeV/c)");
gr[i]->GetXaxis()->SetTitleColor(1);
gr[i]->GetXaxis()->CenterTitle();
gr[i]->GetYaxis()->CenterTitle();
gr[i]->GetYaxis()->SetTitle("#frac{d^{2}N}{dp_{T}dy} (GeV/c)^{-2}");
gr[i]->GetYaxis()->SetTitleOffset(1.28);
gr[i]->SetMarkerSize(1.5);
gr[i]->Fit(fTsallis[i],"REM");
}
// ----------- end of the fiting loop, specify graphs for each particle
gr[0]->SetMarkerColor(kBlue);
gr[0]->SetMarkerStyle(20);
gr[0]->Draw("AP");
gr[1]->SetMarkerColor(kRed);
gr[1]->SetMarkerStyle(21);
gr[1]->Draw("P");
gr[2]->SetMarkerColor(1);
gr[2]->SetMarkerStyle(22);
gr[2]->Draw("P");
TLatex * u = new TLatex();
u->SetTextSize(0.035);
u->SetTextColor(kBlack);
u->DrawLatex(1.5,1.0,"ALICE, pp @ #sqrt{s} = 900 GeV ");
u = new TLatex(0.5,3.0,"#pi^{+}");
u->Draw();
u = new TLatex(0.5,0.3,"K^{+}");
u->Draw();
u = new TLatex(0.5,0.04,"p");
u->Draw();}
```

References

[1] M. D. Azmi et al. "Energy density at kinetic freeze-out in Pb-Pb collisions at the LHC using the Tsallis distribution". In: *Journal of Physics G: Nuclear and Particle Physics* 47 (2020), p. 045001.

[2] T. Bhattacharyya et al. "On the precise determination of the Tsallis parameters in proton–proton collisions at LHC energies". In: *Journal of Physics G: Nuclear and Particle Physics* 45.5 (2018), p. 055001.

[3] J. Cleymans et al. "Hadron resonance gas model and high multiplicities in p–p, p–Pb and Pb–Pb collisions at the LHC". In: *International Journal of Modern Physics E* (2019), p. 1940002.

[4] J. Cleymans et al. "Thermal Model Fits in $p - p$, $p - Pb$ and Pb-Pb Collisions". In: *Journal of Physics: Conference Series*. Vol. 1271. 1. IOP Publishing. 2019, p. 012015.

[5] N. Sharma et al. "Comparison of $p - p$, $p - Pb$, and Pb-Pb collisions in the thermal model: Multiplicity dependence of thermal parameters". In: *Physical Review C* 99.4 (2019), p. 044914.

[6] *Large Hadron Collider*. URL: `https://en.wikipedia.org/wiki/Large_Hadron_Collider`.

[7] S.A Bass. *Virtual Journal on QCD Matter*. 2August 2019. URL: `http://qgp.phy.duke.edu`.

[8] S.A. Bass et al. "Signatures of quark-gluon plasma formation in high energy heavy-ion collisions: a critical review". In: *Journal of Physics G: Nuclear and Particle Physics* 25 (1999), R1.

[9] J.I. Kapusta and E.S. Bowman. "Critical Points in the QCD Phase Diagram with Two Flavors of Quarks". In: *Nuclear Physics A* 830.1-4 (2009), pp. 721c–724c.

[10] J.I. Kapusta and E.S. Bowman. "Multiple critical points in the QCD phase diagram". In: *Arxiv preprint arXiv:0908.0726* (2009).

[11] Vladimir Kekelidze et al. "Status of the NICA project at JINR". In: *EPJ Web Conf.* 138 (2017). Ed. by S. Bondarenko, V. Burov, and A. Malakhov, p. 01027. DOI: `10.1051/epjconf/201713801027`.

[12]   R. Snellings. "Heavy-Ion Collisions: Experimental Highlights". In: *Nuclear Physics A* 820.1-4 (2009), pp. 1c–8c.

[13]   F. Wojciech. *Phenomenology of Ultra-Relativistic Heavy-Ion Collisions*. Clearance Center, 222 Rosewood Drive, Denvers, MA 01923, USA: World Scientific Publishing Co. Pte. Ltd, 2010, pp. 3 –339.

[14]   B. Muller. "Physics and signatures of the quark-gluon plasma". In: *Reports on Progress in Physics* 58 (1995), p. 611.

[15]   M. Csanád and I. Májer. "Equation of state and initial temperature of quark gluon plasma at RHIC". In: *Arxiv preprint arXiv:1101.1279* (2011).

[16]   S. Wheaton, J. Cleymans, and Hauer. M. "THERMUS: A Thermal model package for ROOT". In: *Comput. Phys. Commun.* 180 (2009), pp. 84–106. DOI: 10.1016/j.cpc.2008.08.001. arXiv: hep-ph/0407174 [hep-ph].

[17]   C. Tsallis. "Possible Generalization of Boltzmann-Gibbs Statistics". In: *J. Statist. Phys.* 52 (1988), pp. 479–487. DOI: 10.1007/BF01016429.

[18]   J. Cleymans and D. Worku. "The Tsallis Distribution in Proton-Proton Collisions at $\sqrt{s} = 0.9$ TeV at the LHC". In: *J. Phys.* G39 (2012), p. 025006. DOI: 10.1088/0954-3899/39/2/025006. arXiv: 1110.5526 [hep-ph].

[19]   J. Cleymans and D. Worku. "Relativistic Thermodynamics: Transverse Momentum Distributions in High-Energy Physics". In: *Eur. Phys. J.* A48 (2012), p. 160. DOI: 10.1140/epja/i2012-12160-0. arXiv: 1203.4343 [hep-ph].

[20]   D. Thakur et al. "Indication of a Differential Freeze-out in Proton-Proton and Heavy-Ion Collisions at RHIC and LHC energies". In: *Adv. High Energy Phys.* 2016 (2016), p. 4149352. DOI: 10.1155/2016/4149352. arXiv: 1601.05223 [hep-ph].

[21]   H-L. Lao et al. "Kinetic freeze-out temperatures in central and peripheral collisions: Which one is larger?" In: (2017). arXiv:1703.04944. arXiv: 1703.04944 [nucl-th].

[22]   H-L. Lao, F-H. Liu, and R. A. Lacey. "Extracting kinetic freeze-out temperature and radial flow velocity from an improved Tsallis distribution". In: *Eur. Phys. J.* A53.3 (2017). [Erratum: Eur. Phys. J. A53, no. 6, 143 (2017)], p. 44. DOI: 10.1140/epja/i2017-12333-3, 10.1140/epja/i2017-12238-1. arXiv: 1611.08391 [nucl-th].

[23]   D. Thakur et al. "Indication of Differential Kinetic Freeze-out at RHIC and LHC Energies". In: *Acta Phys. Polon. Supp.* 9 (2016), p. 329. DOI: 10.5506/APhysPolBSupp.9.329. arXiv: 1603.04971 [hep-ph].

[24]   B.I. Abelev et al. "Strange particle production in p+ p collisions at $\sqrt{s} = 200$ GeV". In: *Physical Review C* 75.6 (2007), p. 064901.

[25]   A. Adare et al. "Measurement of neutral mesons in p+p collisions at $\sqrt{(s)}$= 200 GeV and scaling properties of hadron production". In: *Phys. Rev.* D83 (2011), p. 052004. DOI: 10.1103/PhysRevD.83.052004. arXiv: 1005.3674 [hep-ex].

[26]   A. Adare et al. "Identified charged hadron production in $p + p$ collisions at $\sqrt{s} = 200$ and 62.4 GeV". In: *Phys. Rev.* C83 (2011), p. 064903. DOI: 10.1103/PhysRevC.83.064903. arXiv: 1102.0753 [nucl-ex].

[27]   K. Aamodt et al. "Production of pions, kaons and protons in pp collisions at $\sqrt{s} = 900$ GeV with ALICE at the LHC". In: *Eur. Phys. J.* C71 (2011), p. 1655. DOI: 10.1140/epjc/s10052-011-1655-9. arXiv: 1101.4110 [hep-ex].

[28]   V. Khachatryan et al. "Transverse momentum and pseudorapidity distributions of charged hadrons in pp collisions at $\sqrt{s} = 0.9$ and 2.36 TeV". In: *JHEP* 02 (2010), p. 041. DOI: 10.1007/JHEP02(2010)041. arXiv: 1002.0621 [hep-ex].

[29]   V. Khachatryan et al. "Transverse-momentum and pseudorapidity distributions of charged hadrons in pp collisions at $\sqrt{s} = 7$ TeV". In: *Phys. Rev. Lett.* 105 (2010), p. 022002. DOI: 10.1103/PhysRevLett.105.022002. arXiv: 1005.3299 [hep-ex].

[30]   G. Aad et al. "Charged-particle multiplicities in pp interactions measured with the ATLAS detector at the LHC". In: *New J. Phys.* 13 (2011), p. 053033. DOI: 10.1088/1367-2630/13/5/053033. arXiv: 1012.5104 [hep-ex].

[31]   B. B. Abelev et al. "Energy Dependence of the Transverse Momentum Distributions of Charged Particles in pp Collisions Measured by ALICE". In: *Eur. Phys. J.* C73.12 (2013), p. 2662. DOI: 10.1140/epjc/s10052-013-2662-9. arXiv: 1307.1093 [nucl-ex].

[32]   K. K. Gudima et al. "Nuclear Multifragmentation in Nonextensive Statistics: Canonical Formulation". In: *Phys. Rev. Lett.* 85 (22 2000), pp. 4691–4694. DOI: 10.1103/PhysRevLett.85.4691. URL: https://link.aps.org/doi/10.1103/PhysRevLett.85.4691.

[33]   C. Tsallis and Z. G. Arenas. "Nonextensive statistical mechanics and high energy physics". In: *EPJ Web of Conferences*. Vol. 71. EDP Sciences. 2014, p. 00132.

[34] J. Cleymans and M.D. Azmi. "Large transverse momenta and Tsallis thermodynamics". In: *Journal of Physics: Conference Series*. Vol. 668. 1. IOP Publishing. 2016, p. 012050.

[35] G. Wilk and Z. Wlodarczyk. "On the interpretation of nonextensive parameter q in Tsallis statistics and Levy distributions". In: *Phys. Rev. Lett.* 84 (2000), p. 2770. DOI: 10.1103/PhysRevLett.84.2770. arXiv: hep-ph/9908459 [hep-ph].

[36] V. Cirigliano, G. M. Fuller, and A. Vlasenko. "A New Spin on Neutrino Quantum Kinetics". In: *Phys. Lett.* B747 (2015), pp. 27–35. DOI: 10.1016/j.physletb.2015.04.066. arXiv: 1406.5558 [hep-ph].

[37] S. Grigoryan. "Using the Tsallis distribution for hadron spectra in pp collisions: Pions and quarkonia at $\sqrt{s}$ = 5–13000 GeV". In: *Phys. Rev.* D95.5 (2017), p. 056021. DOI: 10.1103/PhysRevD.95.056021. arXiv: 1702.04110 [hep-ph].

[38] C-Y. Wong et al. "From QCD-based hard-scattering to nonextensive statistical mechanical descriptions of transverse momentum spectra in high-energy pp and $p\bar{p}$ collisions". In: *Phys. Rev.* D91.11 (2015), p. 114027. DOI: 10.1103/PhysRevD.91.114027. arXiv: 1505.02022 [hep-ph].

[39] O. Ristea, C. Ristea, and A. Jipa. "Rapidity dependence of charged pion production at relativistic energies using Tsallis statistics". In: *Eur. Phys. J.* A53.5 (2017), p. 91. DOI: 10.1140/epja/i2017-12286-5.

[40] Gábor Bíró et al. "Systematic Analysis of the Non-extensive Statistical Approach in High Energy Particle Collisions - Experiment vs. Theory". In: *Entropy* 19 (2017), p. 88. DOI: 10.3390/e19030088. arXiv: 1702.02842 [hep-ph].

[41] A. S. Parvan. "Ultrarelativistic transverse momentum distribution of the Tsallis statistics". In: *Eur. Phys. J.* A53 (2017), p. 53. DOI: 10.1140/epja/i2017-12242-5. arXiv: 1607.07670 [hep-ph].

[42] A. S. Parvan, O. V. Teryaev, and J. Cleymans. "Systematic Comparison of Tsallis Statistics for Charged Pions Produced In pp Collisions". In: *Eur. Phys. J.* A53.5 (2017), p. 102. DOI: 10.1140/epja/i2017-12301-y. arXiv: 1607.01956 [nucl-th].

[43] S. Tripathy et al. "Nuclear Modification Factor Using Tsallis Non-extensive Statistics". In: *Eur. Phys. J.* A52.9 (2016), p. 289. DOI: 10.1140/epja/i2016-16289-4. arXiv: 1606.06898 [nucl-th].

[44] H. Zheng and L. Zhu. "Comparing the Tsallis Distribution with and without Thermodynamical Description in $p + p$ Collisions". In: *Adv. High Energy Phys.* 2016 (2016), p. 9632126. DOI: 10.1155/2016/9632126. arXiv: 1512.03555 [nucl-th].

[45] L. Marques, J. Cleymans, and A. Deppman. "Description of High-Energy pp Collisions Using Tsallis Thermodynamics: Transverse Momentum and Rapidity Distributions". In: *Phys. Rev.* D91 (2015), p. 054025. DOI: 10.1103/PhysRevD.91.054025. arXiv: 1501.00953 [hep-ph].

[46] M. D. Azmi and J. Cleymans. "The Tsallis Distribution at Large Transverse Momenta". In: *Eur. Phys. J.* C75.9 (2015), p. 430. DOI: 10.1140/epjc/s10052-015-3629-9. arXiv: 1501.07127 [hep-ph].

[47] J. Cleymans et al. "Systematic properties of the Tsallis Distribution: Energy Dependence of Parameters in High-Energy p-p Collisions". In: *Phys. Lett.* B723 (2013), pp. 351–354. DOI: 10.1016/j.physletb.2013.05.029. arXiv: 1302.1970 [hep-ph].

[48] I. Sena and A. Deppman. "Systematic analysis of pT -distributions in p + p collisions". In: *Eur. Phys. J.* A49 (2013), p. 17. DOI: 10.1140/epja/i2013-13017-8. arXiv: 1209.2367 [hep-ex].

[49] M. D. Azmi and J. Cleymans. "Transverse momentum distributions in proton–proton collisions at LHC energies and Tsallis thermodynamics". In: *J. Phys.* G41 (2014), p. 065001. DOI: 10.1088/0954-3899/41/6/065001. arXiv: 1401.4835 [hep-ph].

[50] T. S. Biro and A. Jakovac. "Power-law tails from multiplicative noise". In: *Phys. Rev. Lett.* 94 (2005), p. 132302. DOI: 10.1103/PhysRevLett.94.132302. arXiv: hep-ph/0405202 [hep-ph].

[51] A. Khuntia et al. "Multiplicity Dependence of Non-extensive Parameters for Strange and Multi-Strange Particles in Proton-Proton Collisions at $\sqrt{s} = 7$ TeV at the LHC". In: *Eur. Phys. J.* A53.5 (2017), p. 103. DOI: 10.1140/epja/i2017-12291-8. arXiv: 1702.06885 [hep-ph].

[52] H. Zheng, L. Zhu, and A. Bonasera. "Systematic analysis of hadron spectra in p+p collisions using Tsallis distributions". In: *Phys. Rev.* D92.7 (2015), p. 074009. DOI: 10.1103/PhysRevD.92.074009. arXiv: 1506.03156 [nucl-th].

[53] B. De. "Non-extensive statistics and understanding particle production and kinetic freeze-out process from p_T-spectra at 2.76 TeV". In: *Eur. Phys. J.* A50 (2014), p. 138. DOI: 10.1140/epja/i2014-14138-2. arXiv: 1409.3079 [nucl-th].

[54] C. Tsallis. "Possible generalization of Boltzmann-Gibbs statistics". In: *Journal of Statistical Physics* 52.1 (1988), pp. 479–487. DOI: 10.1007/BF01016429.

[55] J. Cleymans and D. Worku. "The Tsallis distribution in proton–proton collisions at $\sqrt{s} = 0.9$ TeV at the LHC". In: *Journal of Physics G: Nuclear and Particle Physics* 39.2 (2012), p. 025006.

[56] J. Cleymans. "The Tsallis Distribution for p–p collisions at the LHC". In: *Journal of Physics: Conference Series*. Vol. 455. 1. IOP Publishing. 2013, p. 012049.

[57] J. Cleymans. "On the Use of the Tsallis Distribution at LHC Energies." In: *Journal of Physics: Conference Series*. Vol. 779. 1. IOP Publishing. 2017, p. 012079.

[58] J. Cleymans et al. "The Parameters of The Tsallis Distribution at the LHC". In: *EPJ Web of Conferences*. Vol. 137. EDP Sciences. 2017, p. 11004.

[59] L. Altenkämper et al. "Applicability of transverse mass scaling in hadronic collisions at energies available at the CERN Large Hadron Collider". In: *Physical Review C* 96.6 (2017), p. 064907.

[60] R. Witt. "$\langle p(t) \rangle$ systematics and m(t) scaling". In: *Ultra-relativistic nucleus-nucleus collisions. Proceedings, 17th International Conference, Quark Matter 2004, Oakland, USA, January 11-17, 2004*. 2004. arXiv: nucl-ex/0403021 [nucl-ex].

[61] M. Rybczynski and Z. Wlodarczyk. "Tsallis statistics approach to the transverse momentum distributions in p-p collisions". In: *Eur. Phys. J.* C74 (2014), p. 2785. DOI: 10.1140/epjc/s10052-014-2785-7. arXiv: 1401.5639 [hep-ph].

[62] J. D. Bjorken. "Highly Relativistic Nucleus-Nucleus Collisions: The Central Rapidity Region". In: *Phys. Rev.* D27 (1983), pp. 140–151. DOI: 10.1103/PhysRevD.27.140.

[63] M. Ishihara. "Transverse momentum fluctuation under the Tsallis distribution at high energies". In: *International Journal of Modern Physics E* 26.11 (2017), p. 1750071.

[64] R. Brun and F. Rademakers. "ROOT: An object oriented data analysis framework". In: *Nuclear Instruments and Methods in Physics Research* A 389 (1997), pp. 81 –86.

[65] K. Aamodt et al. "Transverse momentum spectra of charged particles in proton-proton collisions at $\sqrt{s} = 900$ GeV with ALICE at the LHC". In: *Phys. Lett.* B693 (2010), pp. 53–68. DOI: 10.1016/j.physletb.2010.08.026. arXiv: 1007.0719 [hep-ex].

[66] B. B. Abelev et al. "Production of charged pions, kaons and protons at large transverse momenta in pp and Pb–Pb collisions at $\sqrt{s_{NN}}$ =2.76 TeV". In: *Phys. Lett.* B736 (2014), pp. 196–207. DOI: 10.1016/j.physletb.2014.07.011. arXiv: 1401.1250 [nucl-ex].

[67] J. Adam et al. "Multiplicity dependence of charged pion, kaon, and (anti)proton production at large transverse momentum in p-Pb collisions at $\sqrt{s_{NN}}$ = 5.02 TeV". In: *Phys. Lett.* B760 (2016), pp. 720–735. DOI: 10.1016/j.physletb.2016.07.050. arXiv: 1601.03658 [nucl-ex].

[68] J. Adam et al. "Measurement of pion, kaon and proton production in proton–proton collisions at $\sqrt{s} = 7$ TeV". In: *Eur. Phys. J.* C75.5 (2015), p. 226. DOI: 10.1140/epjc/s10052-015-3422-9. arXiv: 1504.00024 [nucl-ex].

[69] S. Chatrchyan et al. "Charged particle transverse momentum spectra in *pp* collisions at $\sqrt{s} = 0.9$ and 7 TeV". In: *JHEP* 08 (2011), p. 086. DOI: 10.1007/JHEP08(2011)086. arXiv: 1104.3547 [hep-ex].

[70] A. M. Sirunyan et al. "Measurement of charged pion, kaon, and proton production in proton-proton collisions at $\sqrt{s} = 13$ TeV". In: *Phys. Rev.* D96.11 (2017), p. 112003. DOI: 10.1103/PhysRevD.96.112003. arXiv: 1706.10194 [hep-ex].

[71] S. Chatrchyan et al. "Study of the inclusive production of charged pions, kaons, and protons in *pp* collisions at $\sqrt{s} = 0.9, 2.76,$ and 7 TeV". In: *Eur. Phys. J.* C72 (2012), p. 2164. DOI: 10.1140/epjc/s10052-012-2164-1. arXiv: 1207.4724 [hep-ex].

[72] K. Jiang et al. "Onset of radial flow in p+p collisions". In: *Phys. Rev.* C91.2 (2015), p. 024910. DOI: 10.1103/PhysRevC.91.024910. arXiv: 1312.4230 [nucl-ex].

[73] S. Grigoryan. "Using the Tsallis distribution for hadron spectra in p p collisions: Pions and quarkonia at s= 5- 13000 GeV". In: *Physical Review D* 95.5 (2017), p. 056021.

[74] Z. Tang et al. "Spectra and radial flow at RHIC with Tsallis statistics in a Blast-Wave description". In: *Phys. Rev.* C79 (2009), p. 051901. DOI: 10.1103/PhysRevC.79.051901. arXiv: 0812.1609 [nucl-ex].

[75] F. Cooper and G. Frye. "Single-particle distribution in the hydrodynamic and statistical thermodynamic models of multiparticle production". In: *Phys. Rev. D* 10 (1 1974), pp. 186–189. DOI: 10.1103/PhysRevD.10.186. URL: https://link.aps.org/doi/10.1103/PhysRevD.10.186.

[76] P. F. Kolb and U. W. Heinz. "Hydrodynamic description of ultrarelativistic heavy ion collisions". In: (2003), pp. 634–714. arXiv: nucl-th/0305084 [nucl-th].

[77] J. D. Bjorken. "Highly relativistic nucleus-nucleus collisions: The central rapidity region". In: *Physical review D* 27.1 (1983), p. 140.

[78] A. Erdélyi et al. *Higher transcendental functions, Vol. 1*. McGraw-Hill, New York, 1953.

[79] G. Bíró, G. G. Barnaföldi, and T. S. Biró. "Tsallis-thermometer: a QGP indicator for large and small collisional systems". In: *Journal of Physics G: Nuclear and Particle Physics* (2020). DOI: https://doi.org/10.1088/1361-6471/ab8dcb.

[80] Kea Aamodt et al. "Femtoscopy of p p collisions at s= 0.9 and 7 TeV at the LHC with two-pion Bose-Einstein correlations". In: *Physical Review D* 84.11 (2011), p. 112004.

[81] N. Abgrall et al. "Measurement of negatively charged pion spectra in inelastic p+ p interactions at $p_{lab} = 20, 31, 40, 80$ and 158 GeV/c". In: *The European Physical Journal C* 74.3 (2014), p. 2794.

[82] S. Acharya et al. "Transverse momentum spectra and nuclear modification factors of charged particles in pp, p-Pb and Pb-Pb collisions at the LHC". In: *Journal of High Energy Physics* 2018.11 (2018), p. 13.

[83] C. Alt et al. "Pion and kaon production in central Pb+ Pb collisions at 20 A and 30 A GeV: evidence for the onset of deconfinement". In: *Physical Review C* 77.2 (2008), p. 024903.

[84] S.V. Afanasiev et al. "Energy dependence of pion and kaon production in central Pb+ Pb collisions". In: *Physical Review C* 66.5 (2002), p. 054902.

[85] S. Acharya et al. "Transverse momentum spectra and nuclear modification factors of charged particles in pp, p-Pb and Pb-Pb collisions at the LHC". In: *JHEP* 11 (2018), p. 013. DOI: 10.1007/JHEP11(2018)013. arXiv: 1802.09145 [nucl-ex].

[86] S. Acharya et al. "Transverse momentum spectra and nuclear modification factors of charged particles in Xe-Xe collisions at $\sqrt{s_{NN}}$ =
5.44 TeV". In: *Phys. Lett.* B788 (2019), pp. 166–179. DOI: 10.1016/j.
physletb.2018.10.052. arXiv: 1805.04399 [nucl-ex].

[87] E. Schnedermann, J. Sollfrank, and U. Heinz. "Thermal phenomenology of hadrons from 200A GeV S+S collisions". In: *Phys. Rev. C* 48 (5
1993), pp. 2462–2475. DOI: 10.1103/PhysRevC.48.2462. URL: https:
//link.aps.org/doi/10.1103/PhysRevC.48.2462.

[88] B. I. Abelev et al. "Systematic Measurements of Identified Particle
Spectra in pp, d^+ Au and Au+Au Collisions from STAR". In: *Phys.
Rev.* C79 (2009), p. 034909. DOI: 10.1103/PhysRevC.79.034909. arXiv:
0808.2041 [nucl-ex].

[89] S. Chatterjee et al. "Freeze-out parameters in heavy-ion collisions at
AGS, SPS, RHIC, and LHC energies". In: *Advances in High Energy
Physics* 2015 (2015).

[90] F. Retière and M. A. Lisa. "Observable implications of geometrical and
dynamical aspects of freeze-out in heavy ion collisions". In: *Physical
Review C* 70.4 (2004). ISSN: 1089-490X. DOI: 10.1103/physrevc.70.
044907. URL: http://dx.doi.org/10.1103/PhysRevC.70.044907.

[91] D. Prorok. "Centrality dependence of freeze-out temperature fluctuations in Pb-Pb collisions at the LHC". In: *The European Physical Journal
A* 55.3 (2019), p. 37.

[92] A. Motornenko et al. "Kinetic freeze-out temperature from yields of
short-lived resonances". In: (2019). arXiv: 1908.11730 [hep-ph].

[93] H-T. Ding, F. Karsch, and S. Mukherjee. "Thermodynamics of strong-
interaction matter from Lattice QCD". In: *International Journal of Modern Physics E* 24.10 (2015), p. 1530007.

[94] V. Vovchenko, B. Dönigus, and H. Stoecker. "Canonical statistical model
analysis of $p-p$, p-Pb, and Pb-Pb collisions at energies available at the
CERN Large Hadron Collider". In: *Phys. Rev. C* 100 (5 2019), p. 054906.
DOI: 10.1103/PhysRevC.100.054906. URL: https://link.aps.org/
doi/10.1103/PhysRevC.100.054906.

[95] M. Tanabashi et al. "Review of particle physics". In: *Physical Review D*
98.3 (2018), p. 030001.

[96] H. T. Ding, F. Karsch, and S. Mukherjee. "Thermodynamics of strong-interaction matter from Lattice QCD". In: *Int. J. Mod. Phys.* E24.10 (2015), p. 1530007. DOI: 10.1142/S0218301315300076. arXiv: 1504.05274 [hep-lat].

[97] A. Andronic et al. "Decoding the phase structure of QCD via particle production at high energy". In: *Nature* 561.7723 (2018), pp. 321–330.

[98] R. Stock et al. "The QCD Phase Diagram from Statistical Model Analysis". In: *Nuclear Physics A* 982 (2019), pp. 827–830.

[99] J. Noronha-Hostler and C. Greiner. "Understanding the p/π ratio at LHC due to QCD mass spectrum". In: *Nucl. Phys.* A931 (2014), pp. 1108–1113. DOI: 10.1016/j.nuclphysa.2014.08.101. arXiv: 1408.0761 [nucl-th].

[100] M. Petrán et al. "Hadron production and quark-gluon plasma hadronization in Pb-Pb collisions at $\sqrt{s_{NN}}$ = 2.76 TeV". In: *Phys. Rev.* C88.3 (2013), p. 034907. DOI: 10.1103/PhysRevC.88.034907. arXiv: 1303.2098 [hep-ph].

[101] V. Begun, W. Florkowski, and M. Rybczynski. "Explanation of hadron transverse-momentum spectra in heavy-ion collisions at $\sqrt{s_{NN}} = 2.76$ TeV within chemical non-equilibrium statistical hadronization model". In: *Phys. Rev.* C90.1 (2014), p. 014906. DOI: 10.1103/PhysRevC.90.014906. arXiv: 1312.1487 [nucl-th].

[102] V. Begun, W. Florkowski, and M. Rybczynski. "Transverse-momentum spectra of strange particles produced in Pb+Pb collisions at $\sqrt{s_{NN}} =$ 2.76 TeV in the chemical non-equilibrium model". In: *Phys. Rev.* C90.5 (2014), p. 054912. DOI: 10.1103/PhysRevC.90.054912. arXiv: 1405.7252 [hep-ph].

[103] F. Becattini et al. "Hadron Formation in Relativistic Nuclear Collisions and the QCD Phase Diagram". In: *Phys. Rev. Lett.* 111 (2013), p. 082302. DOI: 10.1103/PhysRevLett.111.082302. arXiv: 1212.2431 [nucl-th].

[104] F. Becattini et al. "Hadronization conditions in relativistic nuclear collisions and the QCD pseudo-critical line". In: *Phys. Lett.* B764 (2017), pp. 241–246. DOI: 10.1016/j.physletb.2016.11.033. arXiv: 1605.09694 [nucl-th].

[105] S. Chatterjee, R. M. Godbole, and S. Gupta. "Strange freezeout". In: *Phys. Lett.* B727 (2013), pp. 554–557. DOI: 10.1016/j.physletb.2013.11.008. arXiv: 1306.2006 [nucl-th].

[106] R. Bellwied et al. "Is there a flavor hierarchy in the deconfinement transition of QCD?" In: *Phys. Rev. Lett.* 111 (2013), p. 202302. DOI: 10. 1103/PhysRevLett.111.202302. arXiv: 1305.6297 [hep-lat].

[107] S. Chatterjee, A. K. Dash, and B. Mohanty. "Contrasting Freezeouts in Large Versus Small Systems". In: *J. Phys.* G44.10 (2017), p. 105106. DOI: 10.1088/1361-6471/aa8857. arXiv: 1608.00643 [nucl-th].

[108] P. Alba et al. "Flavor-dependent eigenvolume interactions in a hadron resonance gas". In: *Nucl. Phys.* A974 (2018), pp. 22–34. DOI: 10.1016/ j.nuclphysa.2018.03.007. arXiv: 1606.06542 [hep-ph].

[109] V. Vovchenko, M. I. Gorenstein, and H. Stoecker. "Finite resonance widths influence the thermal-model description of hadron yields". In: *Phys. Rev.* C98.3 (2018), p. 034906. DOI: 10.1103/PhysRevC.98.034906. arXiv: 1807.02079 [nucl-th].

[110] A. Dash, S. Samanta, and B. Mohanty. "Thermodynamics of a gas of hadrons with attractive and repulsive interactions within an S -matrix formalism". In: *Phys. Rev.* C99.4 (2019), p. 044919. DOI: 10.1103/ PhysRevC.99.044919. arXiv: 1806.02117 [hep-ph].

[111] A. Andronic et al. "The thermal proton yield anomaly in Pb-Pb collisions at the LHC and its resolution". In: *Phys. Lett.* B792 (2019), pp. 304–309. DOI: 10.1016/j.physletb.2019.03.052. arXiv: 1808.03102 [hep-ph].

[112] A. Dash, S. Samanta, and B. Mohanty. "Interacting hadron resonance gas model in the K -matrix formalism". In: *Phys. Rev.* C97.5 (2018), p. 055208. DOI: 10.1103/PhysRevC.97.055208. arXiv: 1802.04998 [nucl-th].

[113] B. I. Abelev et al. "Strange particle production in p+p collisions at $\sqrt{s} = 200$ GeV". In: *Phys. Rev.* C75 (2007), p. 064901. DOI: 10.1103/ PhysRevC.75.064901. arXiv: nucl-ex/0607033 [nucl-ex].

[114] F. Becattini et al. "Predictions of hadron abundances in pp collisions at the LHC". In: *J. Phys.* G38 (2011), p. 025002. DOI: 10.1088/0954-3899/38/2/025002. arXiv: 0912.2855 [hep-ph].

[115] F. Becattini et al. "A Comparative analysis of statistical hadron production". In: *Eur. Phys. J.* C66 (2010), pp. 377–386. DOI: 10.1140/epjc/ s10052-010-1265-y. arXiv: 0911.3026 [hep-ph].

[116] C. Patrignani et al. "Review of Particle Physics". In: *Chin. Phys.* C40.10 (2016), p. 100001. DOI: 10.1088/1674-1137/40/10/100001.

[117] J. Cleymans and H. Satz. "Thermal hadron production in high energy heavy ion collisions". In: *Z. Phys. C - Particles and Fields.* 57 (1993), 135–147. DOI: https://doi.org/10.1007/BF01555746.

[118] J. Cleymans et al. "Comparison of chemical freeze-out criteria in heavy-ion collisions". In: *Phys. Rev. C* 73 (3 2006), p. 034905. DOI: 10.1103/PhysRevC.73.034905. URL: https://link.aps.org/doi/10.1103/PhysRevC.73.034905.

[119] J. Cleymans. "Strangeness: Theoretical status". In: *Proceedings, 3rd International Conference on Physics and astrophysics of quark-gluon plasma (ICPA-QGP '97).* 1998, pp. 55–64. arXiv: nucl-th/9704046 [nucl-th].

[120] W. Broniowski and W. Florkowski. "Explanation of the RHIC p(T) spectra in a thermal model with expansion". In: *Phys. Rev. Lett.* 87 (2001), p. 272302. DOI: 10.1103/PhysRevLett.87.272302. arXiv: nucl-th/0106050 [nucl-th].

[121] S. V. Akkelin, P. Braun-Munzinger, and Yu. M. Sinyukov. "Reconstruction of hadronization stage in Pb + Pb collisions at 158/A-GeV/c". In: *Nucl. Phys.* A710 (2002), pp. 439–465. DOI: 10.1016/S0375-9474(02)01165-X. arXiv: nucl-th/0111050 [nucl-th].

[122] S. Wheaton, J. Cleymans, and M. Hauer. "THERMUS—a thermal model package for ROOT". In: *Computer Physics Communications* 180.1 (2009), pp. 84–106.

[123] P. Braun-Munzinger et al. "Maximum relative strangeness content in heavy ion collisions around 30-GeV/A". In: *Nucl. Phys.* A697 (2002), pp. 902–912. DOI: 10.1016/S0375-9474(01)01257-X. arXiv: hep-ph/0106066 [hep-ph].

[124] J. Letessier et al. "Strangeness conservation in hot nuclear fireballs". In: *Phys. Rev.* D51 (1995), pp. 3408–3435. DOI: 10.1103/PhysRevD.51.3408. arXiv: hep-ph/9212210 [hep-ph].

[125] J. Adam et al. "Enhanced production of multi-strange hadrons in high-multiplicity proton-proton collisions". In: *Nature Phys.* 13 (2017), pp. 535–539. DOI: 10.1038/nphys4111. arXiv: 1606.07424 [nucl-ex].

[126] B. B. Abelev et al. "Multiplicity Dependence of Pion, Kaon, Proton and Lambda Production in p-Pb Collisions at $\sqrt{s_{NN}}$ = 5.02 TeV". In: *Phys. Lett.* B728 (2014), pp. 25–38. DOI: 10.1016/j.physletb.2013.11.020. arXiv: 1307.6796 [nucl-ex].

[127] J. Adam et al. "Multi-strange baryon production in p-Pb collisions at $\sqrt{s_{NN}}$ = 5.02 TeV". In: *Phys. Lett.* B758 (2016), pp. 389–401. DOI: `10.1016/j.physletb.2016.05.027`. arXiv: `1512.07227 [nucl-ex]`.

[128] B. Abelev et al. "Centrality dependence of π, K, p production in Pb-Pb collisions at $\sqrt{s_{NN}}$ = 2.76 TeV". In: *Phys. Rev.* C88 (2013), p. 044910. DOI: `10.1103/PhysRevC.88.044910`. arXiv: `1303.0737 [hep-ex]`.

[129] B. B. Abelev et al. "K^0_S and Λ production in Pb-Pb collisions at $\sqrt{s_{NN}}$ = 2.76 TeV". In: *Phys. Rev. Lett.* 111 (2013), p. 222301. DOI: `10.1103/PhysRevLett.111.222301`. arXiv: `1307.5530 [nucl-ex]`.

[130] B. B. Abelev et al. "Multi-strange baryon production at mid-rapidity in Pb-Pb collisions at $\sqrt{s_{NN}}$ = 2.76 TeV". In: *Phys. Lett.* B728 (2014). [Erratum: Phys. Lett.B734,409(2014)], pp. 216–227. DOI: `10.1016/j.physletb.2014.05.052,10.1016/j.physletb.2013.11.048`. arXiv: `1307.5543 [nucl-ex]`.

[131] *Grace.* 2August 2019. URL: `http://plasma-gate.weizmann.ac.il/Grace/`.

[132] N. Sharma, J. Cleymans, and L. Kumar. "Thermal model description of p–Pb collisions at $\sqrt{s_{NN}}$ = 5.02 TeV". In: *Eur. Phys. J.* C78.4 (2018), p. 288. DOI: `10.1140/epjc/s10052-018-5767-3`. arXiv: `1802.07972 [hep-ph]`.

[133] A. Bazavov et al. "Chiral crossover in QCD at zero and non-zero chemical potentials". In: *Phys. Lett.* B795 (2019), pp. 15–21. DOI: `10.1016/j.physletb.2019.05.013`. arXiv: `1812.08235 [hep-lat]`.

[134] J. Cleymans and L. Kumar. Private Communication. Upper Campus, Rondebosch, Cape Town, 2019.

[135] S. Raghunath. "Relativistic Kinematics". In: *arXiv preprint arXiv:1604.02651* (2016).